Be Your Own
HOUSE
CONTRACTOR

Be Your Own HOUSE CONTRACTOR

Save 25% without Lifting a Hammer

Carl Heldmann

STOREY
BOOKS

*The mission of Storey Publishing is to serve our customers
by publishing practical information that encourages personal
independence in harmony with the environment.*

Edited by Larry Shea
Cover design by Greg Imhoff, modified for this edition
    by Jennifer Jepson Smith
Text design by Cynthia McFarland
Production assistance by Susan Bernier, Beth Morrison,
    and Jennifer Jepson Smith
Indexed by Nan Badgett, Word·a·bil·i·ty

The information in this book is true and complete to the best of our knowledge. All recommendations are made without guarantee on the part of the author or Storey Publishing. The author and publisher disclaim any liability in connection with the use of this information. For additional information, please contact Storey Books, 210 MASS MoCA Way, North Adams, MA 01247.

Storey books are available for special premium and promotional uses and for customized editions. For further information, please call Storey's Custom Publishing Department at (800) 793-9396.

Printed in the United States by R.R. Donnelley
10 9 8 7 6

Library of Congress Cataloging-in-Publication Data

Heldmann, Carl.
   Be your own house contractor: save 25% without lifting a hammer / by Carl Heldmann.
      p. cm.
   Includes index.
   ISBN 1- 58017-374-8 (alk. paper)
   1. House construction–Amateurs' manuals. 2. Building–Superintendence–Amateurs' manuals 3. Contractors–Selection and appointment–Amateurs' manuals. I. Title.

TH4812 .H45 2001
690'.837–dc21
                                                                    2001017023

Contents

Preface to the Fourth Edition vii

Introduction 1

Chapter 1 Be Your Own General Contractor and Save 3

Chapter 2 Where to Start 14

Chapter 3 Financing 30

Chapter 4 Cost Estimating 41

Chapter 5 Further Preparations 52

Chapter 6 Subcontractors 57

Chapter 7 Suppliers 65

Chapter 8 Building the House 69

Chapter 9 Additions to an Existing Home 88

Glossary 90

Appendix 96

Index 131

Acknowledgments

For their tremendous help, I thank my wife, Jane Prante Heldmann, Sharon W. Curran, and Joyce F. Carpenter. My special thanks to Clella Hunt Prante for her invaluable assistance.

Preface

The most expensive words in real estate are "Let's wait." If you wait for interest rates to fall before building a house, you can be sure that the costs of building will increase in the meantime. If you wait for building costs to decrease, it will never happen. **Now** is always the best time to build a house.

But what if you just can't afford "Now"? What do you do? Back in the early 1970s, my wife and I were faced with this very dilemma. We just couldn't afford to build our "dream house" and didn't know what to do. A friend suggested that by building our own home we would save enough money to make the project feasible. I thought he was crazy. "Build my own house?" I said. "I don't know anything about building houses." "Relax," he said, and proceeded to explain to me that by merely acting as the general contractor and hiring all the expert subcontractors (tradespeople), I would save 25 percent or more on the cost of my dream home.

I was skeptical, I admit, but, being the type of person who enjoys finding better ways to solve problems, I decided to try contracting. And the rest, as they say, is history.

I built that first house, saved more than 25 percent, and went on to home building as a career, building millions of dollars worth of homes. Now I am involved in financing new construction. Along the way I wrote this book so that people such as you won't have to make as many mistakes while building their first house as I did. I got through the process of building my first house, and I'm certain that you will as well.

By using this book as a guide, you can avoid many of the pitfalls that plague general contractors, even experienced ones. You will quickly realize that not only will you save 25 percent or more, but by being the "boss" you'll also get exactly the house you want, done the way you want it done. If you want to save a little more money, you can even do some of the labor yourself. More on that later.

This is not a technical book on how to perform the tasks of the various subcontractors you will be hiring. There are plenty of books and videos on carpentry, plumbing, and other specific skilled tasks if you decide you want to get that involved. This is a book that shows you how to be your own general contractor. You will see immediately that you don't have to do any physical labor to save 25 percent or more. You don't even have to lift a hammer. You can hire the pros to do all of that.

I hope that the prospect of being able to save money now, as you build, will enable you to say, "Let's not wait any longer, let's build it **now**." Who knows? Perhaps this venture will launch a new career for you. It certainly did for me. Every professional building general contractor started with his or her first home. Most of us didn't have a book to guide us. You do. Good luck!

Introduction

PEOPLE ALWAYS SEEM to think things are harder than they are. I don't know why this is so, but I'm the same way. Perhaps we fear what we don't know, or maybe it's because someone once told us "It's too hard." Then, when we finally get around to doing it, we discover that the task wasn't as hard as we anticipated.

Everyone who hasn't built a house before thinks it's too hard. For decades, I've been spreading the word that it isn't hard. But I've found that people tend to be skeptical of my enthusiastic belief that "you can do it." However, I've taught thousands of people over the years to do exactly what you are thinking of doing — be your own house contractor. They have come from all walks of life — housewives, salespeople, truck drivers, lawyers, doctors, you name it — and were all motivated by more or less the same reasons. The only variable is the size and style of the house that each one envisioned. And, most important, they all found that they *could* do it and that my enthusiasm was well founded.

Can *you* do it? Is being your own house contractor as easy as I'm making it out to be? Well, I'll tell you a true story that pretty well sums up my answer.

"Is this Carl Heldmann?"

"Yes, it is."

"Are you the one who says it's easy to build your own home?"

By now I was a little nervous and expected to hear the worst. But I said, "I sure am!"

"Well," she replied, "I just wanted to tell you that it's even easier than you said it was!"

I breathed an audible sigh of relief.

The woman went on to say, "My husband and I just finished building our own home in Windy Hill, and we saved so much money

1

and had so much fun that we bought the lot next door and we're going to do it again!"

What more could I possibly add to that testimonial?

You will notice that some technical terms are set in *italics* in this book. These are terms you will need to understand. If you don't, look them up in the Glossary (pages 90–95), where they are defined.

Throughout the book, you will find several sample contracts I have provided to acquaint you with their formats. I urge that you do *not* use these forms unless you have reviewed them with your attorney. State laws differ on what is required in such contracts; your attorney will know what changes may be necessary.

Chapter 1

Be Your Own General Contractor and Save

THIS IS A BOOK that will teach you how to be a general contractor for your own house. There's one big reason for doing it — to save money. How much you will save will vary considerably depending on local prices for labor and materials, land costs, the size of the house, and your ability to follow the steps outlined in this book.

The size of the house will be the largest determining factor, as most general contractors base their profit and overhead on a percentage of the total cost. A larger house will cost more and therefore will include a larger builder's profit and overhead. You will also save more in real estate commissions as the size and value of the house increases.

If you do all that I suggest in this book, a goal of 25 percent savings is possible. If you think of how much you will have to earn — and pay taxes on — to get the money to build, the savings are even greater. And by reducing the amount of money you will need to borrow and then pay back with interest, you will save even more money for years to come.

Let's look at what you can expect to save.

Say that I built a house and I am offering it for sale to the general public. If I set a selling price of $100,000 (a minimum price for a new home in most areas), typically the land would have cost me $25,000, labor and materials $50,000, and my gross profit will be $25,000.

Wow, you say. That's a lot of money to make off one product. Well, if it were that simple, you would be right. But before you get outraged with the building industry in general, let me show you where that gross profit goes when you build professionally.

First, I have to pay sales expense out of my gross profit. If that involves a real estate broker, it may cost as much as $7,000 (7 percent of $100,000). Next, like any business, I have business-expense overhead. This varies with each builder, but the National Association of Home Builders suggests that home builders allocate 50 percent of their adjusted gross profit (after sales expense) for overhead expenses. These include, but are not limited to, phone, insurance, secretarial, gas, truck, car, rent, and office equipment. That leaves me with $9,000, which is half the adjusted gross of $18,000 ($25,000 − $7,000 real estate commission). That is my real expected net (before taxes) profit. You can see that when builders say they make less than 10 percent, they're not wrong.

But you aren't building professionally. You don't have sales expense. You don't have business-expense overhead. You can take the entire gross profit of $25,000 and consider it yours. You may never even have to pay taxes on that gross profit if you follow the Internal Revenue Service guidelines for reinvesting your primary residence capital gains. Imagine! You could actually "make" this kind of money while you go about your regular daily business.

Actual dollar amounts will vary. The average asking price for new homes in many locations may be $150,000 to $200,000, or even more in some areas. But the percentage of actual fair *market value* (what a house should sell for) that you can save will remain fairly true. Does that mean you can expect to save $50,000 on a house that would sell for $200,000? You bet!

What You Need to Know

You need to know very little about building to be a general contractor. You don't need to have the technical knowledge about framing or bricklaying or wiring. Your subcontractors will know their business just as mine do. I'll help you make sure of that.

You may wish to pick up some information on various aspects of building, and that's fine. In most bookstores there are excellent how-to books for the do-it-yourselfer on almost all phases of construction. You may want to read some of them to better understand the process of building a home. But there's no way you can become a master of all trades. The role of the general contractor — which you will be — is as an organizer, a manager, not a tradesman. Your role is to get the job done — by other people.

Hire an Attorney

If, in any of the planning stages of building outlined in this book, you feel uncertain, hire a good real estate attorney. You will need the help of an attorney in certain steps anyway, such as in the closing of the loan. If you hire one now, the fee will be money well spent if it enables you to build your own home, and that fee is an actual cost of construction. Many attorneys will forgo hourly rates in favor of a flat fee, or at least reduce their regular hourly rate if they are going to be involved throughout your project. Be sure that you like the attorney in your first meeting. The two of you will have a long relationship, for the six or more months the house is under construction, so this is important. See page 57 for information on obtaining an attorney.

If you can balance a checkbook, read, and deal with people in a fair manner, you can build your own house. Don't look at building a house as one huge job. View each phase or step as a separate job that you can easily accomplish, and in this way the overall task will seem less monumental.

The most difficult things you will have to do will be behind you when you actually start construction. Sounds unbelievable, doesn't it? But it's true. Your job of planning and organizing will be 90 percent complete when you break ground . . . or it should be. At that point it is up to your team of experts — your subcontractors (subs) — to do its job. Almost all of them will do their jobs correctly, even without your being at the site.

As you go along never be afraid to ask anyone a question about anything. Pride is foolish when it prevents you from asking a question you can learn from.

Carpentry Crew

An important step in building your own home, and maybe even in helping you make the decision to do so, is finding a good carpentry crew. Also known as framing contractors or framers, this carpenter with his crew is the key subcontractor. Having a good one before you start will help you make many intelligent decisions and will help your local lender to make a decision to lend you money.

A personal visit to a few lumberyards or building supply houses (preferably the smaller ones) to ask for recommendations will give you more than enough leads to find a good carpentry crew.

If the first carpenter you find is too busy, ask him to recommend another. But usually another house can be worked into the first carpenter's scheduling. Your carpenter will be one of your best sources for finding many of your other subcontractors.

After selecting a good candidate for this job, check him out. He should give you a list of four or five jobs he's handled. Talk to the

Keeping the Records

Unless you're an exceptionally orderly person, you'll find yourself with construction papers at your desk at home, more in your car glove compartment, and others scattered in pockets or your place of business.

Avoid this by getting organized early. Get a briefcase, which will be your traveling office file. Buy some manila folders, a dozen at least. Then label them as you need them. You'll need one for finances, another for inspections, with a schedule of inspections made out by you, one for each of the subs — and more.

There are two reasons for doing this. First, it will keep your work flowing more smoothly if you don't have to paw through papers to find what you want. Second, there's a feeling of confidence that you will get just from seeing your part of the job arranged in such a businesslike way.

owners. Look at the work he did for them. Others to check with are banks and building suppliers.

One note about this key subcontractor: Most good carpenters, particularly the older ones — mid-thirties and up — have a very good knowledge of most phases of construction. Though they may not be up-to-date on wiring for Internet access or the latest energy-efficient appliances, these job-site veterans are experts on getting a house framed up and closed in quickly, with all doorways, stairways, and window openings sized and located correctly. The good carpenters are easy to find, hard to get, and usually worth waiting for. They will also cost you a bit more but, in the long run, will save you money and time.

Time Involved

How much time is involved in being your own contractor? For each individual it will be different, but you'll spend more hours planning and getting ready than you will on the construction site — that is, if you let the subs do their jobs without feeling that you must be there every minute.

In the planning stages you may spend a month of your spare time or several months, depending on how fast you are able to make a decision on all the various aspects that I will cover. For example, deciding on house plans can take an hour, a month, or even longer; the same thing is true when you are choosing land, specifications, and subcontractors.

The time involved after you start construction will be short. The maximum time ought not to exceed two hours a day; those hours need not interfere with your job or normal activities. You will have many helpers and some free management. You can have real estate people (the ones you are dealing with on your lot) do some legwork by checking on restrictions and getting information on septic systems and wells. Your suppliers can help you save time, too, by finding your subs for you, doing *"take-offs"* for estimating and ordering supplies, and giving technical advice.

Building your own home is an ideal husband/wife operation. Just be sure that you and your spouse agree and get your signals straight. Failure to do this has often been a source of petty annoyance for me with some of my clients. Problems arise because of poor communication.

Get What You Want

This reminds me of one of the main benefits of building your own house. You will get more of what you want in your house, with fewer hassles, when you act as your own general contractor. Most contractors are very staid. They like to do things the same old way. This often causes problems for the buyer who wants something done a little differently. With you as the general contractor, you can get things done as differently as you like.

As a contractor, the success of your house-building venture depends on good communication. Most established subcontractors have cell phones and/or pagers, which can make communication a lot easier. To ensure better communication, one very simple and not too costly step you can take is to put a phone on the building site. Long distance calls are easily controlled. In my many years of contracting I've never had the job phone abused by a sub. It has saved me countless trips to the job site, too. Just be sure you or a sub takes the phone home each night.

Years ago I built a vacation home in the mountains about 140 miles from our permanent residence. I made only three trips to supervise and check on the progress the entire time the house was being built, from the *staking* of the lot to the final inside trim. I don't recommend this. I only mention it to show that you don't have to be there every minute or every day. My mountain house was built by phone. This was far cheaper and much less tiring than driving 140 miles each way. The subs did a beautiful job. They wouldn't have done any better if I had been there watching them.

Work by Phone

Phone calls to subs are the key to this business. Make those calls before they leave for work, or in the evening when you are not working. You can make on-site inspections before or after work or on your lunch hour. Daily on-site inspections usually aren't necessary. If you believe they are, and you can't do it, your spouse or a friend can. You don't have to be there every minute. There isn't a general contractor alive who is. For me that would mean that I could build only one house at a time, but when I was building, I usually had five or six under construction at once.

You don't have to watch masons lay every brick or your carpenters pound every nail. You allow them to do their jobs. Mistakes may be made from time to time. Chances are they would have been made if you were there. There isn't a mistake in the whole process that can't be rectified. It should be your individual decision on how much time you want to be at the site.

You Can Get a Loan

Unless you are paying cash for your house, you will need a loan. Almost all loans for houses are made by banks, mortgage companies, and credit unions.

For all practical purposes, getting your loan is the most important step, and the most crucial in building. No money, no house. You must determine very early in your decision-making process whether you will be able to get a loan.

The intricacies of financing are explained in Chapter 3, where I discuss how a loan works and how much you can borrow. I will be very honest and say that most lenders will be somewhat reluctant to lend to an individual who plans to be his or her own general contractor. This reluctance is legitimate. Lenders want to be sure the house will be built properly and, most important, finished. They don't want to step in and finish a house. They are lenders, not builders, and that's the way they would like to keep it. It will be a challenging sales job convincing them you can be your own general contractor, but with the knowledge you will receive from this book, you can do it.

Plan your visit carefully. You want to appear to be a person who has thought through all steps in the construction of the house, who understands the problems and has solutions to them. You know the lender won't be enthusiastic about your decision to be your own contractor, so you must do all in your power to present this approach as an asset, not a liability, and your projected savings can make it an asset.

The lender will wonder whether you are an organized person who can handle the details of home construction. You will be convincing if you have all of your homework completed when you reach the bank. "Make it look as though the only step left is to get the money," suggested one lender.

Another Lender's Views

Lenders, I've learned, don't differ too much across the country. They tend to be wary when you mention being your own general contractor. They're willing to be cooperative if they think you have a good chance of finishing the job, but they don't want to be stuck with a half-completed house.

A New England lender quickly rattled off the three requirements necessary for any borrower, covering credit references, a financial statement, and proof of income.

I asked what would make him more warmly disposed toward the would-be general contractor. He listed these three points:

◆ Own your lot, free and clear.
◆ Show proof of your ability to handle the construction, as described in this book.
◆ Have as much planning completed as possible to assure the lender that this is not an idle dream, but a project that lacks only the needed financing. Such planning would include a set of house plans, tentative arrangements with suppliers (including estimates of the costs of the materials), and all information needed on such subjects as availability of utilities, any zoning restrictions, and local building regulations.

Information You Need

Your presentation should include a package of material, including the following:

1. Proof of your current income. Include joint incomes from all sources.
2. Good credit references.
3. A financial statement of what you own and what you owe.
4. Evidence that you have sufficient knowledge about being your own general contractor.

Numbers 1, 2, and 3 are prerequisites to buying or building under any circumstances. Number 4 is what this book is all about. The lender will check out your knowledge of the topics covered in all of the chapters and will easily recognize your ability to estimate, find a lot, and select plans, and your knowledge of construction loans, permanent mortgages, subcontractors, suppliers, and the different phases of construction. Don't worry — all of these items are explained in this book.

Don't Be Discouraged

After you become familiar with the material in this book, there should be no reason for you not to get a loan. However, you should keep in mind that you may get turned down by one or two lenders before you get an approval. Don't be discouraged. Sometimes it is nothing more than personal chemistry, or other trivial things, and has nothing to do with your ability.

It costs nothing to go through the initial stages of discussion with a lender, and you may get a commitment for a loan. In building my first house, I was turned down by four lending institutions before I found two who would permit me to be my own general contractor. If I hadn't been persistent, who knows where I would be right now? I was turned down because I didn't know everything you'll know when you finish reading this book. I didn't show sufficient knowledge until after the third rejection. Each time I was turned down, I would do a little more homework and gain more confidence in myself. I didn't have a book — you do. You can do it.

An Alternative, the Manager's Contract

When you decide to act as your own general contractor you must do a good selling job to the lender. And, after you've given your best effort, let's suppose the worst happens. You can't get a loan with yourself as the general contractor. Or, let's suppose you simply don't feel able to deal with subs.

Here are three other ways you can build, each using a general contractor in a position of increasing responsibility and at greater cost to you:

1. A manager's contract.
2. Cost plus a percentage or a fixed fee contract.
3. Contract bid.

These are listed in the order that they usually increase costs to you. Because each one increases the contractor's responsibility, the cost of having that contractor will also increase. The cost of the materials to build the house, the land, and other fixed expenses should remain the same.

For all practical purposes, options 2 and 3 are readily available contracts and are used by most builders with varying degrees of legalese and varying degrees of slant favoring one or the other party. (Samples of contracts for all three options are shown in the Appendix on pages 96 to 105.)

The least expensive way to go — and one by which you can still be considered the general contractor, be accepted by your lender, and possibly feel more comfortable dealing with subs — is the first option, the *manager's contract*.

Under this arrangement, you have a licensed general contractor who has one responsibility, to act as your manager with the subs. This contractor, in return for about one-third to two-thirds of the normal fee or profit and overhead, will assist in finding the subs (although you can still find your own), schedule subs, check the quality of their work, approve the quality of materials, and order materials, when needed, in your name. You will be responsible for securing suppliers, permits, loans; paying all bills, including those from subs; and ensuring inspections for quality and approval. You will be responsible for the final job and its overall acceptance.

Some of the advantages of this contract are that the general contractor will:

◆ Do a cost estimate.

◆ Help find subs.

◆ Schedule your time better.

◆ Help cut red tape and get the necessary permits.

◆ Arrange for temporary services.

◆ Schedule subs, from survey to landscaping.

◆ Develop materials lists.

◆ Assist in reviewing bills when requested.

A slight change from this procedure would be to hire a designer-builder who would advise you on plans, make any necessary modifications to those plans, review the specifications for such things as adequate insulation or the possible use of native materials, and inspect the process of construction.

In either case, the manager's contract shown in the Appendix could be modified to include more or fewer of the responsibilities of the general contractor. The cost, of course, will vary with the number of responsibilities listed.

If you decide to use one of the contracts in the Appendix, go over it with your attorney to check it for applicability and to make certain it conforms with local laws.

Chapter 2

Where to Start

Y OU'VE MADE YOUR decision to build. Wonderful. You'll do well, so don't worry. What comes next? Well, where do you want to live, based on your needs and desires and finances?

Remember in making this decision that you can use this house as a stepping-stone. Once you've built your own home, you can move up to a bigger or better house or location and possibly at no — and I mean no — extra cost. Each time you move you are adding to your profit as a builder and thus are increasing your own *equity*.

Land

If you have lived in a city or town for a while, you should know where you want to build. If you haven't looked around, or if you are new to an area, I suggest using a local real estate broker. These brokers know what each area offers and what lots cost in different locations. If acreage is what you are after, here, too, a real estate broker is most helpful. Don't forget, your broker is one of those helpers I mentioned. A broker can help you locate the property you want and can help with all the details necessary to assure you that it is a suitable building site. The broker should be able to show you a map of the lot and point out the boundaries to you during a walk around the lot.

If you don't already own land, first choose the area where you want to live.

Next, look for a lot or a site with enough acreage in that area, one that is suitable to build on.

Do You Need a Basement?

Do you want — or need — a basement? Here are the pros and cons on building a house with a basement.

Pros:

◇ It's relatively inexpensive, compared with aboveground space, particularly if you must dig down 6 or more feet anyway for placing *footings* and foundation walls, as you must in some parts of the country.

◇ A basement gets your heating system, and sometimes the fuel supply, out of the way.

◇ It's a good storage space — but it may be too damp for paper, metal, and clothing.

◇ On a sloped lot, a basement may provide you with garage space.

◇ It's an excellent space for an out-of-the-way family room.

Cons:

◇ If the water table is high, flooding may be an annual headache.

◇ A basement may be too damp for many uses, or require a dehumidifier.

◇ It's an ideal spot for storing things better taken to the dump.

◇ Building a basement can be expensive, particularly if the excavator encounters ledge (large rock mass) that requires blasting or a spring that keeps the hole flooded.

Sloping Lots

If you want a basement and you live in an area of the country where the soil doesn't drain well, avoid flat lots. A sloping lot will provide drainage by means of footing drains. Another advantage of a sloping lot is that it allows you to design your house with a walk-out basement with doorways or windows on one side to provide natural light and ventilation. Also, in some areas the open side can be *frame construction,* which is a little less expensive than poured concrete or concrete block.

If you don't want a basement, try to find a relatively flat lot so that you won't have an excessive amount of *crawl space* (or fill, if you decide on a slab foundation). Crawl space is cheaper than a basement. If the lot slopes on only one end or corner, though, it will not cost too much more to accommodate the foundation to the lot. Notice that you are accommodating the foundation to the lot, not the other way around. I can't say it strongly enough: Find the land first and then make the plans to fit it.

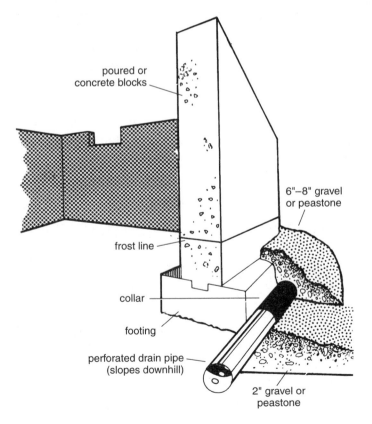

A footing drain will help to keep your cellar dry.

Trees Are Valuable

If you like trees, try to find a wooded lot within your budget. Barren lots are cheaper to build on but more costly to landscape — almost an exact cost trade-off. A wooded lot generally costs more to buy, more to build on, and less to landscape. It's your choice, but if you are concerned about top-dollar resale in the near future, it's worth putting a few thousand dollars extra into a wooded lot.

If the area in which you are looking has no development activity near it, I strongly recommend *test borings* of the soil — *before* you purchase the lot — to determine its *load-bearing capability*. These tests will also show whether there is ledge on the site that might require blasting. The test is not expensive, and the seller should be willing to pay for it. Without a soil test, you can end up paying thousands of extra dollars for foundation and drainage work. Be sure that this test is included as a contingency in any contract to buy. It can also be included as a refund provision in the contract in the event that non–load-bearing soil is discovered after purchase.

Specialists in test boring for load bearing are listed in the Yellow Pages under "Engineers, Consulting" or "Engineers, Foundation." Most county health departments make soil tests for septic systems. This is done for free or at a very low cost. These departments also provide information about wells.

Let the Brokers Do the Work

Put the burden of getting these tests on your real estate brokers. Let them do the legwork and checking. Just make sure it's in the contract you're offered. Let them earn their commissions. Let them hand you a nice, clean, finished deal — a lot ready for you to build your dream house on.

When you deal with a realtor — a member of the National Association of Realtors — if there is any problem you have a recourse, and that is the association. In almost all towns and cities, the local Board of Realtors is so image conscious and worried about a member ruining that image that it will protect you, the buyer. Simply call the board if you believe a realtor isn't doing a good job or isn't representing your best interests.

Water and Sewage

You will need to consider both water and sewage as you select your building lot.

If water and sewage are both provided by your city or town, you need only find out how to tap into them, and what are the costs involved.

If either one is provided (and quite often it is only water), you will not face much of a problem. You may need some form of a permit, plus a percolation test for a septic system, to determine whether the soil will absorb the discharge of the system.

If you must provide both water and sewage, some study of the situation is needed.

A septic system consists of a sewage pipe running from your home to an underground septic tank, usually made of concrete. Sewage enters the tank and is broken down by bacterial action. A liquid then flows from the tank into a system of perforated pipes that distribute it into an area called a leach field. The size of the field required depends on the amount of sewage going into the tank and the ability of the soil to absorb

Typical layout for a septic system. The leach field must be located in an area where the soil can absorb moisture.

moisture. Sandy soil is best because of its permeability; denser, claylike soil can create problems.

Tougher regulations in many communities may require that new septic systems be designed by an engineer. In many areas, regulations require that a well be a certain distance, say a minimum of 150 feet, from your leach field or from the field of any neighbor. Thus it's best to have a large lot if you must provide both water and a sewage system.

Digging a well can be expensive. You can get a dowser, complete with forked branch, to "locate" water for you, or you can do what most well diggers do, and that is to locate the well where it's most convenient, and start digging.

Most well diggers charge by the feet of depth that they bore, and they usually can estimate about how far they will have to dig. The depth of existing wells in the area is sometimes (but not always) a good indication of how deep you'll have to drill to reach water, so try to use a well driller who is familiar with the area where your lot is located. Some diggers will offer you a contract price for the well.

well casing

water level

centrifugal pump

pump intake

motor

A common type of installation in a home well. The pump and motor are positioned at the bottom of the well.

Zoning

This is the most important question to ask when deciding to buy a building site: Is it a buildable lot? Your local building inspection department is responsible for answering this question. If they will issue a building permit it can be considered a buildable lot. Be sure to check with your local building inspection department before you purchase your building site.

Consider zoning carefully when you are choosing a lot or acreage. Be aware of what could later be built in your future neighborhood — stores, offices, trailer parks, or industry. Look at what is already there. Such things as dumps, railroads, and industrial buildings can make a lot worth much less as a site for a home. One could write a book on zoning — and many have. A quick consultation with your real estate broker and your attorney should resolve any fears you have.

If water and sewer are provided, be sure your real estate broker (or you or your attorney, if no broker is used) checks out all costs to get the water and sewer to your property line if it is not already there (it may be across the street). You also need to find out about any tap-in fees or privilege fees charged by the municipality or association providing the services. Do the same for gas, electrical, and phone services.

Thoroughly investigating wells, the septic system, load-bearing capabilities, water and sewer fees, and zoning may take a few weeks. If you like the lot and don't want to risk losing it, you can make an offer with the understanding that your binder or deposit will be refunded if the conditions mentioned above are not in your favor. Don't put more than 10 percent down as a binder or deposit, and if you are using a broker, be sure that the money is held in *escrow*. Above all, consult your attorney before closing a deal on any piece of land. I could say that a thousand times, and it wouldn't be enough. Have him or her check the title to the lot to make sure it is clear. Check out all restrictions on the size and type of dwelling you can build, and what a neighbor may build. These restrictions may or may not be specifically covered in local zoning regulations. Different parts of the country use different means to protect an area from visual blight. Again, your broker and your attorney can help you on this.

How Much Should You Spend?

Land costs probably vary more than any other item in the construction business. Three things determine the selling price of land. They are location, location, and location. I'm sure you've heard that before, but it's true. However, the more you spend on land, the less you'll have left over for the house.

You will need to determine building costs in your area, plan a total budget, and then shop for land based on what you have left after subtracting building costs. Building costs can be quickly determined in any area of the country by using my "windshield estimating"

Checklist for Buying a Lot

Here are some of the many things you will want to check before buying a lot:

Building permit
   Most important: Can you get one for the site?_____
Neighborhood
   Will your house fit in with others on street?_____
   Quality of schools?_____
   Attractive street?_____
Transportation
   Public transportation available?_____
Pollution (air, water, or noise, such as airport)_____
Zoning restrictions_____
Sewage, water available_____
Soil tests made_____
Direction of slope, if any_____
Drainage (Observe after heavy rain, if possible.)_____
Test borings_____
Size of lot_____
Trees (valuable, but sun needed for garden)_____
Title clear_____
Cost (A counteroffer of 10–20 percent below asking price
        is often accepted.)_____

method described in the estimating chapter. If your total budget (cash + mortgage) is, for example, $100,000 and you find a piece of property that sells for $40,000, and local building costs are $50 per square foot (with you as the general contractor), that means you could only build a 1,200-square-foot home. $100,000 – $40,000 = $60,000 ÷ $50 (per square foot) = 1,200 square feet. That's not a very big house for such an expensive piece of property.

It's a juggling process between land cost and house size, but ideally, the land should be worth approximately 25 percent of the fair market value of the house and land together.

Buying land is a very subjective process. One tip I'll pass on, however, is to always keep resale value in mind. I know that may not seem important to you, especially since selling is the farthest thing from your mind right now, but some day it won't be. It is always on the mind of your mortgage lender.

The House Plans

There are several ways of getting house plans. One is to find a plan suitable for you and your building site by looking through magazines or books of plans. House plans are also available on the Internet and on CD-ROMS. There are computer programs that allow you to design your own house plans, although these are not always as easy to use as the creators of these programs would lead you to believe. Another way to obtain plans is to have them drawn for you by an architect or drafter/designer.

The least expensive way to obtain plans is through the thousands of magazines, books, and CD-ROMs of plans that you will find. Quite often you'll find that one of the plans needs only minor modifications to make it suitable for you.

You either should stick with the plans the way they are or order the minimum number sold (sometimes this is only one set) and get advice from a drafter/designer or an architect on any changes, no matter how minor. This will be much less expensive than having the entire plan drawn for you. An experienced draftsman can even make major changes in a set of plans and advise you on the practicality of changes you suggest and the additional cost, if any, of those changes.

Make Changes Early

Try to make all changes in the plans before you start construction. You need not have the plans redrawn for minor changes such as moving, adding, or deleting a window or door. But you should if you move walls or change roof lines and *roof pitches.*

If the drafter/designer redraws the plans, it will cost about twenty to thirty cents per square foot for a finished product, including six to eight sets of the plans. If he or she designed your house from scratch, you could expect to pay from thirty to fifty cents per square foot. In that case you usually would get more precisely what you wanted.

You could hire an architect to do this work. This would cost considerably more. Either a drafter/designer or an architect can do the job and either one should give you a *quote* over the phone of approximate costs. Look for names in the Yellow Pages or get recommendations from friends. Shop for price. Before you select an individual or firm, ask to see work done, and be sure to ask for a few references. Don't be timid and, most of all, don't be intimidated. After all, you will be the employer.

Cutting Construction Costs

Your first chance to save a lot of money is now, when you are selecting the plans for your home.

◇ Do you need that many square feet? Every reduction you can make in size will cut the costs.

◇ One story or two? It's cheaper to build the same number of square feet of housing in a two-story house.

◇ What is its shape? A simple rectangle is the cheapest to build. Curves, extra corners, and extensions in the floor plan will all increase costs.

◇ What style roof? The least expensive is the common gable roof. Dormers and multiple gables make the roof more difficult to build and therefore more costly.

◇ How's your timing? If you're building when there's a lull in construction (such as in winter), chances are you'll get your money's worth — and maybe a bit more.

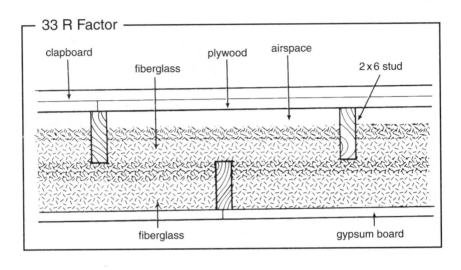

33 R Factor
clapboard   fiberglass   plywood   airspace   2 x 6 stud
fiberglass   gypsum board

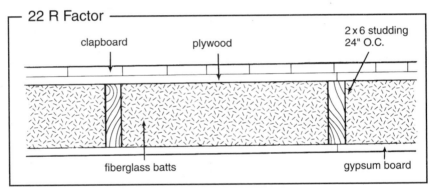

22 R Factor
2 x 6 studding 24" O.C.
clapboard   plywood
fiberglass batts   gypsum board

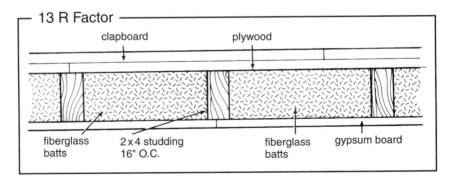

13 R Factor
clapboard   plywood
fiberglass batts   2 x 4 studding 16" O.C.   fiberglass batts   gypsum board

Three ways (and there are many more) of building outer walls, and their R factors. At top, this wall has an R factor of about 33, with the 2 x 6 studs not touching both the inner and outer layers. Center, with 2 x 6 studding, is about R-22. At bottom is about R-13, the conventional method used until a few years ago.

On-the-job charges are expensive, so make your decisions on paper and live with them or expect high cost overruns. I usually rough sketch my houses on quarter-inch graph paper and make most of my changes in this stage. If you plan to start from scratch with your plans, you can do this too. A 12-by-12-foot room will be drawn

Insulation

You're lucky. While those of us who own homes built before 1970 wish we had a chance to increase the insulation in them, you are starting fresh and can plan a home that is energy efficient.

Consider this as you select your plan and as you talk over your plans with your subs, particularly your carpenter and your insulation sub. There are decisions you can make that will result in thousands of dollars of savings as you live in your home over the years. Some decisions will require changes in your house plans and specifications.

If you have a full basement, should you insulate it? Do your windows provide two layers of glass? How will your outer walls be built? (Three possibilities are shown on page 24.) How much insulation will you have over your head? Will your heating ducts and hot-water pipes be insulated? Is your chimney inside the house, where any heat it collects will be returned to the house, or on an outer wall, where heat will be lost? Does your fireplace provide a source of outside air so that you're not sending heated air up the chimney, and does it have a tight damper so hot air isn't flowing up from it when it isn't in use? A wood-burning stove, a pellet stove, or a gas fireplace can give you much of the ambience of a traditional fireplace while saving you hundreds of dollars in construction costs and hundreds more later by providing more efficient, less expensive heating.

You'll find utility companies discuss R factors. This refers to the ability of a material to resist the escape of heat. A high R factor means that you are not losing as much heat. Recommended minimum R factors for walls and roofs vary geographically; check with your building inspector to see how much insulation is typically used in your area.

three squares by three squares. This doesn't allow for wall thickness, and I don't worry about it in the early planning stages unless there's a tight squeeze with a room or a stairway and I need to know then that it's going to fit. If you don't feel confident doing this kind of planning, don't do it. Buy your plans or have them drawn for you from the sketching stage up. I only do it because I enjoy it. It really is a job for one of your subs, your architect or drafter/designer.

Stay closely involved with this phase, even if you don't sketch the plan. Pay strict attention to what is in the floor plans. Picture yourself walking through the house from room to room. If you are unsure as to whether a room is large enough, find a room of about the same size, and compare.

This is a good time to mention the only two tools you'll need as a general contractor: two steel tape measures, one 25 feet long and the other 100 feet. You may not use them many times, but they will help you eliminate guesswork in the planning stages.

Studying Plans

Figuring out the plans is easy, although it may look difficult at first. You'll be looking first at room sizes, room placement, traffic flow, kitchen work flow, closet space, number of baths, and overall size. Those are the major functions of design. Without too much effort you can move doors and windows, add them, or delete them — on paper.

Unless you have an unlimited budget, and very few of us do, a house is a series of compromises. We can't afford all of those beautiful things we see in home magazines. A house is going to cost so much per square foot using the average number of windows and doors and average appliances. If you want a $4,000 Jacuzzi bath and can't afford to spend $4,000 more, compromise on something else — the size of the house, the garage, or a paved driveway.

Don't make any of your decisions a "matter of life and death." You may be surprised how unimportant that "big decision" was when you are finished. Agonizing over decisions only leads to friction — within yourself and with others. Make all of your decisions conscientiously but quickly, and then move on.

Be sure your plans include specifications (specs). Specifications are lists of materials that are going into your house. Most plans come with a set of specs. Your drafter or architect will include a set with the drawings.

There will be a "for input" in the specs for you. Refer to the Description of Materials form on page 106 in the Appendix, and you will see that there are places for indicating decorative items, such as wallpaper, carpet, paint, stain, and door hardware. Obviously, you will choose these. Since specs come early in the game, most people haven't selected these decorative items at that point. Therefore, monetary allowances are used with the term "or equal." This means that the actual item finally selected, such as a kitchen faucet, will be of an approximately equal dollar amount and of similar quality to the one in the specs.

Parts of Plans

Be sure your plans include the following pages, all of which are shown here or in the Appendix.

◊ **A plot plan.** If your plans are drawn locally, the designer can prepare this plan for you. Otherwise you or a surveyor will have to do it. It is merely a plat (map) of your lot with the position of the house drawn or blocked in. Its purpose is to ensure that a given house plan will fit on a given lot. If it doesn't fit on paper, it surely won't fit in reality.

First, the lot is drawn, then all the setbacks required by zoning and/or restrictions are sketched in, then the location of the house is indicated. The exact location may be slightly different when the house is physically staked out, but will be close to the indicated position on the plat — especially if it is a tight fit. If you have a lot consisting of several acres, the actual position of the house from that indicated on the plat could be changed by many feet.

The plot plan won't be drawn first if you have a set of plans but no lot. It's better to find a lot first, then make the plans to fit. That is not an absolute rule. Some people love a set of plans or they want a particular house they have seen in a magazine. They would rather search for a lot to fit the plans. It can be done, although not always easily or inexpensively.

Plot plan

◇ Spec sheets. In these lists of materials for your house, spell out as much detail as possible. Using an existing set (such as those found in the Appendix on pages 106–16) is the easiest way to start. Make any necessary modifications.

◇ Foundation plan. This shows the overall dimensions of the house and the locations of all load-bearing requirements, such as *piers*, steel *reinforcing rods*, vents, basement slabs, and, if a basement house, all walls, windows, doors, and plumbing of that basement. (In the Appendix you will find a foundation plan.)

◇ **Floor plans.** You must have a plan for each floor. It will
show outside dimensions plus locations of windows and
doors, plumbing fixtures, and large appliances. You may add
the electrical drawings as well. These drawings show the
locations of receptacles and switches. It is not necessary to
have these shown on the floor plans. After a house is
framed, you can go through it with an electrician and locate
receptacles, light fixtures, switches, cable outlets, and
phone lines. However, when getting quotes or bids from
subs, it is best to have the electrical, cable, and phone
requirements included either in the floor plans or on a separate sheet. The same is true for heating and cooling
(mechanical) requirements. It is also best to have these
drawn out in advance, for bid purposes and to locate the furnace. I use *heat pumps* exclusively in my houses so I don't
have to worry about furnace placement. If you plan to heat
with wood, coal, gas, or oil, then planning the *flue* location is
necessary, especially in a house without a basement and/or
in a two-story house, unless your appliance is one of the
newer direct-vent models.

◇ **Detail sheet.** This is a sheet showing cabinet details, cross
sections of the inside of the house if there is a section that
may not be perfectly clear from the floor plan, and wall sections to show all the materials that make up a wall, such as
brick, sheathing, studs, insulation, and inside wallboard or
paneling. A detailed section of a foundation wall is also
sketched in order to show proper construction technique,
drainage, and *waterproofing*.

◇ **Outside elevations.** These are the sheets (usually two) that
show an outside view of how the house will look when finished. Usually all four sides are shown.

You'll need about six sets of plans: one for yourself, one for your
lender, and one for each of the major subs. If you can get sets of
plans that are inexpensive, it's good to have as many as possible. If
you're going out for bids, for example, for work by a sub, each of
the bidders should have a copy of the plans to study.

Chapter 3
Financing

IF YOU FOLLOWED the instructions in Chapter 1, by now you have convinced a bank, mortgage company, or credit union to give you a loan. That loan will be divided into two phases.

1. The construction loan.
2. The permanent mortgage loan, to be paid off in 15 to 30 years.

Usually any difficulties in obtaining a loan involve the first phase, the construction loan, not the permanent loan given when the house is finished.

The Construction Loan

If you were able to secure a $90,000 mortgage for the construction of your new house and the lender were to give you a check for that amount before you started, both you and the lender would be in trouble. The lender, because it just loaned $90,000 on a house that doesn't yet exist; you, because you'd have to pay interest and principal on $90,000 before you even broke ground for that new house.

Considering interest expense, that could mean you would have to pay a large amount of interest each month on a nonexistent house, along with your regular monthly fixed expenses for items such as food and rent.

Enter the construction phase of mortgage loans — the construction loan. When you are approved for a mortgage for a house yet to be built, the lender will set up your loan first on a short-term basis, called the construction loan. The money is disbursed to you as the house progresses.

Only Interest Paid

You pay only interest, no principal, on the total amount lent to you for any given month during construction. For example, if the house is 20 percent complete after the first month, the lender, after a physical inspection of the progress of construction, will advance 20 percent of your total to you to pay your bills. The schedule on page 32 shows how this percentage is determined. If your total mortgage is, say, $90,000, then you will receive $18,000.

The house is now 20 percent complete, you are one month along in progress, and you have paid no interest yet. And you won't have to until next month, when you will then pay the interest, say at an annual 10 percent rate for 30 days, on the $18,000 you received. That payment will be $150, which you can pay out of the $18,000.

This is extremely important. Many people are afraid they will face two house payments at the same time. Not so. You use dollars that were advanced for construction to pay the monthly interest. You make no repayment of principal. You do not use your money. Construction loan interest is a cost of construction of any new home, whether you build it or I build it for you. It's not even part of the builder's overhead. It's a cost like lumber and nails. Treat it as such and plan for it, along with other financing charges, and it won't bother you. Think of all the money you are saving by being your own general contractor.

There is a considerable time period during construction in which the house is progressing toward completion, but you haven't yet received bills for materials. It will take you approximately six months to complete the house, and because of the billing time lag from suppliers and loan interest being one month or more in arrears, during that six months you will have had an average of only 40 percent of the total money available disbursed to you. This is an approximate figure, obviously, as is the amount of time needed to

INSPECTION REPORT AND DISBURSEMENT SCHEDULE

Borrower _____

Contractor _____

Job address _____

City _____

	% OF TOTAL	MONTHLY INSPECTIONS				
		1	2	3	4	5
1. Excavation	2					
2. Foundation & slab work	8					
3. Floor	4					
4. Wall framing	6					
5. Roof framing & sheeting	6					
6. Wall sheathing	1					
7. Roofing	2					
8. Windows set	10					
9. Exterior doors	2					
10. Siding/brick	7					
11. Exterior trim	1					
12. Well/water tap	2					
13. Septic sewer tap	2					
14. Plumbing roughed	5					
15. Wiring roughed	2					
16. Heating roughed	2					
17. Heat plant	2					
18. Insulation	2					
19. Chimney/flue	2					
20. Drywall	6					
21. Bath tile	1					
22. Plumbing fixtures	4					
23. Interior trim	3					
24. Cabinets	3					
25. Exterior painting	2					
26. Interior painting	2					
27. Appliances	2					
28. Light fixtures	1					
29. Carpet/floor finish	4					
30. Drives and walks	3					
31. Landscaping	1					
TOTAL	100%					

Inspector _____

Comments: _____

complete your house. I'm merely pointing out that you won't pay interest on the full amount of your mortgage until the house is complete. At that time the lender will complete the disbursement of the construction money and convert the loan to a permanent mortgage for 15 to 30 years. Then you will be making your full payment of principal and interest. But then you have your dream house. Move in. If that sounds simple, it is only because it is.

Two Homes?

By the way, even in a sluggish market, houses sell sooner or later. If you already have a home and you fear it won't sell before your new one is ready, either don't worry about it or sell it first and live in an apartment while building your new home.

The other financing charges will vary with lenders and attorneys but can be estimated in advance. They will total about 2 to 3 percent of the mortgage amount. They may have to be paid when you close your construction loan, or they may be deducted from the total construction loan amount, such as the $90,000 above.

I prefer the latter method. It doesn't tie up my cash. The additional financing costs are also a construction cost.

Cash Needed

You will need some extra cash. Call it *interim* (short-term) *financing.* Some expenses will occur before your lender makes the first disbursement of construction money, and you will need a few thousand dollars to bridge that gap for such things as paying subs, fees, and permits. If you have cash available, fine. If not, you may want to borrow from a commercial bank or take a second mortgage on your existing home, or a *bridge loan.*

Excluding the cost of the land and the 2 to 3 percent in financing charges I discussed, you shouldn't need more than an additional 5 percent of the cost of the house for interim operating expenses. This money is needed only until the house is finished. When the lender disburses all the money in the form of the permanent loan, the need for your operating cash will disappear and you can put it back where you got it.

If it is borrowed, include the interest paid in your cost estimate. It, too, is a cost of construction that other builders and I have included. It may seem that I put the cart before the horse by discussing interest charges first, but many people worry about them and I want to allay that worry.

FHA and VA loans are government backed and deal *only* with permanent financing and not construction financing. Your local lender will coordinate the construction loan.

A Choice of Mortgages

Here is a short description of some of the mortgages currently available.

Conventional mortgage. My favorite because it involves the least amount of red tape and usually is fastest to secure. You may have to have a larger *down payment*, depending on the laws in your state and the requirements of the lender.

Federal Housing Administration Mortgage. FHA insures the loan. The down payments required are small, and the repayment period is up to 35 years, but it may take time for you to get your loan approved and for FHA to check out your plans. Your lender will have specific details.

Veterans Administration Mortgage. You have to qualify as a veteran to get the VA to guarantee your loan. No down payment is required in some cases. Approval is sometimes slow. Ask your lender for specific details.

Farmers Service Agency Mortgage. Both guaranteed mortgages and direct loans are available from this agency of the U.S. Department of Agriculture. (This program was formerly run by the Farmers Home Administration agency.) The interest rate you pay depends on your family income and the size of your family. This can be an excellent way to

Usually the lenders will require that your land be paid for in full and that you have a *clear title* to it before they will lend you money to build on it. This is to protect them. Quite simply put, if you don't have clear title to the land, they won't have clear title to their *note* to you for construction money.

We don't need to go into these matters any further here, as your attorney can explain the legalese far better. It is not necessary for you to become an expert on the law to be your own general contractor.

borrow for a low-income family. Check this one by telephoning the USDA, listed in your telephone directory under U.S. Government, Agriculture Department.

A number of other ways to finance homes have developed since the late 1970s, when interest rates were high. Depending on the state of the economy when you are seeking a mortgage, there may well be additional options you will want to investigate.

The most simple is the **mortgage financed by the seller.** Used more often for home sales than for lot sales, this mortgage has terms negotiated by the two parties involved.

The **graduated payment mortgage** is tailored for young people with expectations of higher incomes. Monthly payments are small at first but gradually increase. Figure out total payments under this mortgage and compare them with those of a conventional mortgage before being swayed by those smaller early monthly payments.

The **variable-rate mortgage** can be written so that monthly payments or the total number of those payments can be changed during the life of the mortgage as interest rates move up or down.

Lot Subordination

There is a way to avoid paying for the land before construction: *lot subordination*. It means the owner will take a note in lieu of payment of all or part of the asking price and legally subordinate his or her interest in that land to your lender. Consult your attorney for details. Some lenders won't allow this or any other form of borrowing for the land or down payment. Usually the land is the down payment, as its value is close to the normal dollar amount put down on a house. Lenders want that money to be your equity. They may have you sign a statement that you aren't borrowing your down payment.

But in your case, you are saving an amount equal to your down payment or your land by being your own builder. The builder's profit and overhead you are saving is a legitimate figure. You will earn it and it will be an actual dollar amount when the house is finished.

Lenders will have different opinions of many of your particular financing needs. You will need to shop around and spend some time talking to them. They will help you tremendously in putting your financial package together. In this aspect, they are like subcontractors. Each of them will also explain its policies on construction draws, inspections, construction loan interest payments, *points*, credit reports, qualifications, and interest rates.

The Permanent Mortgage

Once you've secured funds for construction, there are several things to keep in mind as you seek a permanent mortgage.

Qualifying for a Loan

As to qualifications or "qualifying for a loan," the requirements change daily and are different for each lender. A rule of thumb that is often bent is that your annual mortgage payments (only on your permanent mortgage) should not exceed 2 to 2½ times your gross monthly income from all sources and from both spouses. Local lenders will have various rules, but most will be about like this. Also, the more money you put down (increasing your equity), the more a lender will be willing to lend, and often at a lower interest rate.

Don't forget, the thousands of dollars you are saving by acting as general contractor are a legitimate value and should be, under all circumstances, considered as cash equity.

Keep this in mind when talking to lenders: Many people are squeezed out of the housing market by rising prices, and even more are unable to buy a home when interest rates go up. Yet lenders still must make loans to stay in business, as it is a major source of their income. Perhaps you might have been squeezed out in the past yourself, but now, by being your own general contractor, you can handle a loan. The difference could be that money you'll save by going this route.

How Big a Mortgage?

How much house you need and therefore how much you need to borrow is up to you. Chapter 4 offers a fast, easy, and fairly accurate method of estimating the costs. Stretching your budget for more house is not all bad, because:

◆ You get a house large enough for a growing family.

◆ You are making a good investment. Houses appreciate in value while you get an excellent tax write-off on the tax-deductible interest.

◆ Your income will probably catch up in a few years. Check into graduated mortgage payments or variable interest rates.

◆ You can always bail out by selling the completed house and recoup all costs plus a profit. Then you're ready, of course, to reinvest in another new home.

Financing may be the most difficult phase of construction as it will be the one in which you may feel the least adequate. You will find, however, that after talking to one or two lenders, you'll be more relaxed and better versed in their jargon. Of course, you may get lucky and establish a good relationship with the first lender you approach.

STANDARD MORTGAGE APPLICATION FORM

Uniform Residential Loan Application

This application is designed to be completed by the applicant(s) with the lender's assistance. Applicants should complete this form as "Borrower" or "Co-Borrower", as applicable. Co-Borrower information must also be provided (and the appropriate box checked) when ☐ the income or assets of a person other than the "Borrower" (including the Borrower's spouse) will be used as a basis for loan qualification or ☐ the income or assets of the Borrower's spouse will not be used as a basis for loan qualification, but his or her liabilities must be considered because the Borrower resides in a community property state, the security property is located in a community property state, or the Borrower is relying on other property located in a community property state as a basis for repayment of the loan.

I. TYPE OF MORTGAGE AND TERMS OF LOAN

Mortgage Applied for:	☐ VA ☐ FHA	☐ Conventional ☐ FmHA	☐ Other:	Agency Case Number		Lender Case No.
Amount $	Interest Rate %	No. of Months	Amortization Type:	☐ Fixed Rate ☐ GPM	☐ Other (explain): ☐ ARM (type):	

II. PROPERTY INFORMATION AND PURPOSE OF LOAN

Subject Property Address (street, city, state, & zip code)		No. of Units
Legal Description of Subject Property (attach description if necessary)		Year Built

Purpose of Loan	☐ Purchase ☐ Refinance	☐ Construction ☐ Construction-Permanent	☐ Other (explain):	Property will be: ☐ Primary Residence ☐ Secondary Residence ☐ Investment

Complete this line if construction or construction-permanent loan.

Year Lot Acquired	Original Cost	Amount Existing Liens	(a) Present Value of Lot	(b) Cost of Improvements	Total (a + b)
	$	$	$	$	$

Complete this line if this is a refinance loan.

Year Acquired	Original Cost	Amount Existing Liens	Purpose of Refinance		Describe Improvements ☐ made ☐ to be made
	$	$			Cost: $

Title will be held in what Name(s)	Manner in which Title will be held	Estate will be held in: ☐ Fee Simple ☐ Leasehold (show expiration date)
Source of Down Payment, Settlement Charges and/or Subordinate Financing (explain)		

III. BORROWER INFORMATION

Borrower / Co-Borrower

Borrower's Name (include Jr. or Sr. if applicable)	Co-Borrower's Name (include Jr. or Sr. if applicable)

Social Security Number	Home Phone (incl. area code)	Age	Yrs. School	Social Security Number	Home Phone (incl. area code)	Age	Yrs. School

☐ Married ☐ Unmarried (include single, divorced, widowed) ☐ Separated	Dependents (not listed by Co-Borrower) no. / ages	☐ Married ☐ Unmarried (include single, divorced, widowed) ☐ Separated	Dependents (not listed by Borrower) no. / ages

Present Address (street, city, state, zip code) ☐ Own ☐ Rent _____ No. Yrs.	Present Address (street, city, state, zip code) ☐ Own ☐ Rent _____ No. Yrs.

If residing at present address for less than two years, complete the following:

Former Address (street, city, state, zip code) ☐ Own ☐ Rent _____ No. Yrs.	Former Address (street, city, state, zip code) ☐ Own ☐ Rent _____ No. Yrs.

Former Address (street, city, state, zip code) ☐ Own ☐ Rent _____ No. Yrs.	Former Address (street, city, state, zip code) ☐ Own ☐ Rent _____ No. Yrs.

IV. EMPLOYMENT INFORMATION

Borrower / Co-Borrower

Name & Address of Employer	☐ Self Employed	Yrs. on this job	Name & Address of Employer	☐ Self Employed	Yrs. on this job
		Yrs. employed in this line of work/profession			Yrs. employed in this line of work/profession
Position/Title/Type of Business	Business Phone (incl. area code)		Position/Title/Type of Business	Business Phone (incl. area code)	

If employed in current position for less than two years or if currently employed in more than one position, complete the following:

Name & Address of Employer	☐ Self Employed	Dates (from - to)	Name & Address of Employer	☐ Self Employed	Dates (from - to)
		Monthly Income $			Monthly Income $
Position/Title/Type of Business	Business Phone (incl. area code)		Position/Title/Type of Business	Business Phone (incl. area code)	
Name & Address of Employer	☐ Self Employed	Dates (from - to)	Name & Address of Employer	☐ Self Employed	Dates (from - to)
		Monthly Income $			Monthly Income $
Position/Title/Type of Business	Business Phone (incl. area code)		Position/Title/Type of Business	Business Phone (incl. area code)	

V. MONTHLY INCOME AND COMBINED HOUSING EXPENSE INFORMATION

Gross Monthly Income	Borrower	Co-Borrower	Total	Combined Monthly Housing Expense	Present	Proposed
Base Empl. Income*	$	$	$	Rent	$	
Overtime				First Mortgage (P&I)		$
Bonuses				Other Financing (P&I)		
Commissions				Hazard Insurance		
Dividends/Interest				Real Estate Taxes		
Net Rental Income				Mortgage Insurance		
Other (before completing, see the notice in "describe other income," below)				Homeowner Assn. Dues		
				Other:		
Total	$	$	$	Total	$	$

* Self Employed Borrower(s) may be required to provide additional documentation such as tax returns and financial statements.

Describe Other Income *Notice:* Alimony, child support, or separate maintenance income need not be revealed if the Borrower (B) or Co-Borrower (C) does not choose to have it considered for repaying this loan.

B/C		Monthly Amount
		$

VI. ASSETS AND LIABILITIES

This Statement and any applicable supporting schedules may be completed jointly by both married and unmarried Co-Borrowers if their assets and liabilities are sufficiently joined so that the Statement can be meaningfully and fairly presented on a combined basis; otherwise separate Statements and Schedules are required. If the Co-Borrower section was completed about a spouse, this Statement and supporting schedules must be completed about that spouse also. Completed ☐ Jointly ☐ Not Jointly

ASSETS Description	Cash or Market Value	Liabilities and Pledged Assets. List the creditor's name, address and account number for all outstanding debts, including automobile loans, revolving charge accounts, real estate loans, alimony, child support, stock pledges, etc. Use continuation sheet, if necessary. Indicate by (*) those liabilities which will be satisfied upon sale of real estate owned or upon refinancing of the subject property.		
		LIABILITIES	Monthly Payt. & Mos. Left to Pay	Unpaid Balance
Cash deposit toward purchase held by:	$	Name and address of Company	$ Payt./Mos.	$
List checking and savings accounts below				
Name and address of Bank, S&L, or Credit Union				
		Acct. no.		
		Name and address of Company	$ Payt./Mos.	$
Acct. no.	$			
Name and address of Bank, S&L, or Credit Union				
		Acct. no.		
		Name and address of Company	$ Payt./Mos.	$
Acct. no.	$			
Name and address of Bank, S&L, or Credit Union				
		Acct. no.		
		Name and address of Company	$ Payt./Mos.	$
Acct. no.	$			
Name and address of Bank, S&L, or Credit Union				
		Acct. no.		
		Name and address of Company	$ Payt./Mos.	$
Acct. no.	$			
Stocks & Bonds (Company name/number & description)	$			
		Acct. no.		
		Name and address of Company	$ Payt./Mos.	$
Life insurance net cash value	$			
Face amount: $				
Subtotal Liquid Assets	$			
Real estate owned (enter market value from schedule of real estate owned)	$	Acct. no.		
Vested interest in retirement fund	$	Name and address of Company	$ Payt./Mos.	$
Net worth of business(es) owned (attach financial statement)	$			
Automobiles owned (make and year)	$			
		Acct. no.		
		Alimony/Child Support/Separate Maintenance Payments Owed to:	$	
Other Assets (itemize)	$	Job Related Expense (child care, union dues, etc.)	$	
		Total Monthly Payments	$	
Total Assets a.	$	Net Worth (a minus b) ➡	$	Total Liabilities b. $

VI. ASSETS AND LIABILITIES (cont.)

Schedule of Real Estate Owned (If additional properties are owned, use continuation sheet.)

Property Address (enter S if sold, PS if pending sale or R if rental being held for income) ↓	Type of Property	Present Market Value	Amount of Mortgages & Liens	Gross Rental Income	Mortgage Payments	Insurance, Maintenance, Taxes & Misc.	Net Rental Income
		$	$	$	$	$	$
	Totals	$	$	$	$	$	$

List any additional names under which credit has previously been received and indicate appropriate creditor name(s) and account number(s):

Alternate Name	Creditor Name	Account Number

VII. DETAILS OF TRANSACTION

a. Purchase price	$
b. Alterations, improvements, repairs	
c. Land (if acquired separately)	
d. Refinance (incl. debts to be paid off)	
e. Estimated prepaid items	
f. Estimated closing costs	
g. PMI, MIP, Funding Fee	
h. Discount (if Borrower will pay)	
i. Total costs (add items a through h)	
j. Subordinate financing	
k. Borrower's closing costs paid by Seller	
l. Other Credits (explain)	
m. Loan amount (exclude PMI, MIP, Funding Fee financed)	
n. PMI, MIP, Funding Fee financed	
o. Loan amount (add m & n)	
p. Cash from/to Borrower (subtract j, k, l & o from i)	

VIII. DECLARATIONS

If you answer "yes" to any questions a through i, please use continuation sheet for explanation.

	Borrower		Co-Borrower	
	Yes	No	Yes	No
a. Are there any outstanding judgments against you?	☐	☐	☐	☐
b. Have you been declared bankrupt within the past 7 years?	☐	☐	☐	☐
c. Have you had property foreclosed upon or given title or deed in lieu thereof in the last 7 years?	☐	☐	☐	☐
d. Are you a party to a law suit?	☐	☐	☐	☐
e. Have you directly or indirectly been obligated on any loan which resulted in foreclosure, transfer of title in lieu of foreclosure, or judgment? (This would include such loans as home mortgage loans, SBA loans, home improvement loans, educational loans, manufactured (mobile) home loans, any mortgage, financial obligation, bond, or loan guarantee. If "Yes," provide details, including date, name and address of Lender, FHA or VA case number, if any, and reasons for the action.)	☐	☐	☐	☐
f. Are you presently delinquent or in default on any Federal debt or any other loan, mortgage, financial obligation, bond, or loan guarantee? If "Yes," give details as described in the preceding question.	☐	☐	☐	☐
g. Are you obligated to pay alimony, child support, or separate maintenance?	☐	☐	☐	☐
h. Is any part of the down payment borrowed?	☐	☐	☐	☐
i. Are you a co-maker or endorser on a note?	☐	☐	☐	☐
j. Are you a U.S. citizen?	☐	☐	☐	☐
k. Are you a permanent resident alien?	☐	☐	☐	☐
l. Do you intend to occupy the property as your primary residence? If "Yes," complete question m below.	☐	☐	☐	☐
m. Have you had an ownership interest in a property in the last three years?	☐	☐	☐	☐
(1) What type of property did you own--principal residence (PR), second home (SH), or investment property (IP)?				
(2) How did you hold title to the home--solely by yourself (S), jointly with your spouse (SP), or jointly with another person (O)?				

IX. ACKNOWLEDGMENT AND AGREEMENT

The undersigned specifically acknowledge(s) and agree(s) that: (1) the loan requested by this application will be secured by a mortgage or deed of trust on the property described herein; (2) the property will not be used for any illegal or prohibited purpose or use; (3) all statements made in this application are made for the purpose of obtaining the loan indicated herein; (4) occupation of the property will be as indicated above; (5) verification or reverification of any information contained in the application may be made at any time by the Lender, its agents, successors and assigns, either directly or through a credit reporting agency, from any source named in this application, and the original copy of this application will be retained by the Lender, even if the loan is not approved; (6) the Lender, its agents, successors and assigns will rely on the information contained in the application and I/we have a continuing obligation to amend and/or supplement the information provided in this application if any of the material facts which I/we have represented herein should change prior to closing; (7) in the event my/our payments on the loan indicated in this application become delinquent, the Lender, its agents, successors and assigns, may, in addition to all their other rights and remedies. report my/our name(s) and account information to a credit reporting agency; (8) ownership of the loan may be transferred to successor or assign of the Lender without notice to me and/or the administration of the loan account may be transferred to an agent, successor or assign of the Lender with prior notice to me; (9) the Lender, its agents, successors and assigns make no representations or warranties, express or implied, to the Borrower(s) regarding the property, the condition of the property, or the value of the property.

Certification: I/We certify that the information provided in this application is true and correct as of the date set forth opposite my/our signature(s) on this application and acknowledge my/our understanding that any intentional or negligent misrepresentation(s) of the information contained in this application may result in civil liability and/or criminal penalties including, but not limited to, fine or imprisonment or both under the provisions of Title 18, United States Code, Section 1001, et seq. and liability for monetary damages to the Lender, its agents, successors and assigns, insurers and any other person who may suffer any loss due to reliance upon any misrepresentation which I/we have made on this application.

Borrower's Signature	Date	Co-Borrower's Signature	Date
X		X	

X. INFORMATION FOR GOVERNMENT MONITORING PURPOSES

The following information is requested by the Federal Government for certain types of loans related to a dwelling, in order to monitor the Lender's compliance with equal credit opportunity, fair housing and home mortgage disclosure laws. You are not required to furnish this information, but are encouraged to do so. The law provides that a Lender may neither discriminate on the basis of this information, nor on whether you choose to furnish it. However, if you choose not to furnish it, under Federal regulations this Lender is required to note race and sex on the basis of visual observation or surname. If you do not wish to furnish the information, please check the box below. (Lender must review the above material to assure that the disclosures satisfy all requirements to which the Lender is subject under applicable state law for the particular type of loan applied for.)

BORROWER ☐ I do not wish to furnish this information

Race/National Origin: ☐ American Indian or Alaskan Native ☐ Asian or Pacific Islander ☐ Black, not of Hispanic origin ☐ Hispanic ☐ White, not of Hispanic origin ☐ Other (specify) _____

Sex: ☐ Female ☐ Male

CO-BORROWER ☐ I do not wish to furnish this information

Race/National Origin: ☐ American Indian or Alaskan Native ☐ Asian or Pacific Islander ☐ Black, not of Hispanic origin ☐ Hispanic ☐ White, not of Hispanic origin ☐ Other (specify) _____

Sex: ☐ Female ☐ Male

To be Completed by Interviewer This application was taken by:	Interviewer's Name (print or type)	Name and Address of Interviewer's Employer
☐ face-to-face interview	Interviewer's Signature Date	
☐ by mail		
☐ by telephone	Interviewer's Phone Number (incl. area code)	

Chapter 4
Cost Estimating

Estimating the cost of a new house is not an exact science. You can, however, come reasonably close to estimating a total cost, and you can guide your construction project toward that estimated cost. You can do this only if *you* are the general contractor.

There is a catch-22 in building a home with regard to estimating costs. As you'll see, you can't get a really accurate estimate of costs until you have house plans, and you can't — or shouldn't — get house plans until you know how much a certain style and size home will cost to build in your area of the country. If you spend the money to buy plans and then find that it's too expensive to build that particular plan, you've wasted time and money.

Well, fear not, for I've invented a process for determining how much it costs to build a house anywhere in the United States or Canada. I call this process "windshield estimating." It is called this because you can literally determine building costs from the comfort of your car as you drive around your area and find the style and size home you're interested in. Visit areas where quality custom homes are being built by small to medium-sized builders, for this is your ultimate goal — a quality custom home contracted by a very small builder (you). Find a style of home you like and a size you find adequate. Don't worry if the selling prices of the houses you are looking at seem too high for your budget, as your only concern is to determine building costs.

Let's say you find a nice two-story house with a three-car garage and a full, unfinished basement listed for sale at $XX.00. All you have to do then is subtract 25 percent of the selling price, then subtract the property cost, and you are left with the actual cost of building that house. Divide that final cost figure by the square footage of living space, and you've got the cost per square foot in your area. Simple? You bet! Does this really work? Yes, for most areas. Just be sure you look at as many houses as you can to get a good feel for what is accurate and fair pricing in your area. If you live in a rural area and not much is being built, your strategy will need to change somewhat. Travel to the nearest town, buy the newspaper to check out home prices, or consult with a local realtor.

Now, there are two questions that should have come to mind immediately when you read the above. One, how do you know the property cost? And two, how do you know the square footage of the house you're looking at so that you can divide and get the cost per square foot? In answer to question one, the property cost can be determined by checking the selling price of adjoining property or nearby comparable property, contacting a realtor, or calling the local tax office. (All real estate transactions and records are always a matter of public record.) You might also consider the direct approach: Check with the builder.

The builder is the most obvious source of information, as he or she will be interested in your future business and would therefore be willing to tell you not only the lot value, but — guess what! — the square footage as well. If you can't talk to the builder to find out square footage, call the building inspection department and ask (as it too is a matter of public record) the square footage that was recorded when the building permit was issued. At any rate, with a bit of perseverance, you'll come up with building costs on a square foot basis without a lot of effort.

Now you can determine what style and size house plan to start looking for. And, once you get your plans, you can then do a more precise estimate. For more precise estimating, draw up an estimate sheet like the one on page 45 and fill in each item based on bids from subcontractors and suppliers. Enter these bids under the "estimated cost" column. The "actual cost" column is for entering paid bills. This second column will be accurate because it will be filled in when each category is completed and has been paid for.

Most lumber companies will do a take-off (estimate) of the materials they hope to sell you. For your first house I'd advise you to let them. That doesn't mean you have to buy from them. More on this in Chapter 7. Your carpenter can also do a take-off of materials he will use. He may or may not charge for this service. Ask him.

As you pay for items during construction, compare them to your original estimate so you can see how costs are running. If costs run higher than estimated in some categories, you may want to trim some costs in others.

Most of the heavy costs come at the beginning of construction for such items as excavation work, well and septic digging and installation, lumber, masonry, carpentry, plumbing, electrical work, windows and doors, and HVAC (heating, ventilating, and air-conditioning). Some of the items bought later still have a very significant affect on the total cost of the house. These include hardwood floors and other flooring, appliances, plumbing fixtures, wallpapering, millwork and interior trim, and carpeting. It is possible to cut costs on many of these items by choosing less expensive options or postponing installation of certain items (such as wallpaper) until after you've moved in.

I have seen houses with $5,000 worth of *molding* just on the first floor. Molding and trim details (such as a wainscot or cornice molding) can be added later if reducing immediate costs is important. Windows can vary as much as $100 each just because of brand name, and the difference in quality is negligible. Do you need a ten-cycle dishwasher, when a five-cycle or a two-cycle will do as well? So it is with each item.

Balancing Costs

Use a form such as the one on page 45 to estimate the cost of each construction item and to later fill in the actual costs. If you figure $8,000 for framing lumber and supplies, and the actual cost turns out to be $8,500, start looking for other places to trim. If the actual cost is $7,500, don't become too elated and start looking for ways to spend that $500. It will disappear before you finish, I promise.

Why would a single item vary so much? For three reasons:

1. Prices of materials vary daily — sometimes substantially.
2. Different tradespeople use different amounts of materials. Some carpenters overbuild and will use $400 to $500 worth

of additional lumber to do the same job as someone else. (Let them — I'd rather overbuild in framing than underbuild.) This is just one house, and you can save somewhere else. If you were building ten to fifty houses a year, you would need to be more careful with materials used.

3. It is impossible to determine the exact number of *studs*, *joists*, bricks, and other materials that go into a house. Some estimators are more accurate than others, but an exact estimate is pure luck.

4. For many carpenters, it's standard procedure to order more lumber than is necessary (up to 5 percent or so) to have enough on hand to replace any boards that may be too badly warped, bowed, or split to use.

Don't worry yourself to death over the lack of exactness. A builder pads an estimate by at least 5 percent to cover these contingencies. That is just one reason why you'll be saving by being your own general contractor. Don't worry about minor variations in amounts of materials used, but keep an eye on each category as you get bids and prices.

The number of bricks and concrete blocks required is the most difficult to estimate. Brick companies often will estimate for you, but they won't be any more accurate than you on your first try.

You have a choice of making a rough estimate and ordering from it, or having the brick company do it. If you don't have enough bricks, you can order more. If you order too many, you can return the surplus. (Ask the brick company about its return policy when you're shopping for your brick.) Or, quite often, your masonry sub will buy unused brick at a discounted price and haul them off the job for you. Ordering bricks may be the most difficult part of estimating materials — but only if you let it bother you. I recommend ordering a little short and letting the brick company hold some extra bricks in its brickyard in your name so that if you need to order more, you'll get the same *color run*.

You'll need about five to seven bags of *mortar* cement per 1,000 bricks. Sand is ordered in such enormous quantities (14 or 16 cubic yards) that you don't have to concern yourself with exactness. If I seem to oversimplify a very complex item — good. Don't worry

COST ESTIMATE

Owner's name _____

Property address _____

City _____

	ESTIMATED COST	ACTUAL COST
Permits, fees, surveys, etc.	_____	_____
Utilities (electric, gas, phone)	_____	_____
Excavation	_____	_____
Foundation	_____	_____
Rough Lumber	_____	_____
Rough Labor	_____	_____
Windows & Exterior Doors	_____	_____
Roofing	_____	_____
Concrete Flatwork	_____	_____
Siding	_____	_____
Plumbing	_____	_____
Heating	_____	_____
Electrical	_____	_____
Insulation	_____	_____
Water (Well)	_____	_____
Sewer (Septic)	_____	_____
Fireplaces	_____	_____
Drywall	_____	_____
Cabinets	_____	_____
Interior Trim	_____	_____
Interior Trim Labor	_____	_____
Painting	_____	_____
Appliances	_____	_____
Light Fixtures	_____	_____
Floor Coverings	_____	_____
Driveway	_____	_____
Garage Door	_____	_____
Other	_____	_____
Subtotal	_____	_____
Misc. (5% to 10% recommended)	_____	_____
Building Cost	_____	_____
Land Cost	_____	_____
Total (Land & Building) Cost	_____	_____

Down Payment _____ **Construction Loan** _____

about it. Order a little less than you think you need of sand, brick, and block, and if you run out, order more. Your masonry sub will be sure to let you know in advance if he is going to run short of an item, because if he doesn't he will be held up.

Many masonry subs can estimate their needs without your help. If you find one who does, let him. Some even supply their own materials — better yet.

Never be too proud to ask questions, from either a supplier or subcontractor. Both are very willing to help. They don't make any money until they sell you something or perform a service. Also, some plans purchased in magazines come with — or make available for a fee — a schedule of materials, right down to the amount of nails. They are fairly accurate, and I wouldn't hesitate to use them. Suppliers also do this for free. (See Chapter 7 for types of suppliers to approach.)

Estimating Bricks Needed

To figure the area of a wall, multiply the length times the width (or the base times the height). If the wall is to be 8 feet high and the base is 20 feet long, the area is 160 square feet.

If you plan a brick foundation one *course* in thickness, which is what you would have on a house with a relatively low crawl space, here is how to determine how many standard 2" x 4" x 8" bricks you would need.

The foundation is 30' x 60' x 3'. Treat each wall as a separate area, determine that area, add all areas together, and multiply by 8. Multiplying by 8 is generous because each square foot of surface area requires 7.5 bricks plus mortar space.

So our 30' x 60' x 3' foundation would use the following:

Side 1	30 x 3 = 90 square feet
Side 2	60 x 3 = 180 square feet
Side 3	30 x 3 = 90 square feet
Side 4	60 x 3 = 180 square feet
Total	540 square feet x 8 = 4,320 bricks

Cost Breakdowns

Here are the items and categories you will most likely encounter in building your home. With some, accurate bids from suppliers or sub-contractors are possible, but with others, a guesstimated allowance can be used.

- ◇ **Permits, fees, surveys.** Your building inspection department can give you the cost of permits and fees, which vary by locale. Your lender will require you to have a mortgage survey, and you will also have to insure your building project against loss due to fire or other acts of God. Any other costs necessary to get your project started can be included in this category and some guesstimating can be utilized.

- ◇ **Utilities (electric, gas, phone).** In some rural areas this item can be very costly, running into thousands of dollars. Check with your local utility companies in advance.

- ◇ **Excavation.** This item will depend on such factors as locale, soil conditions, terrain, and season of the year. Excavating for a basement will cost more than excavating for a slab or a crawl space. Get a written bid if possible.

- ◇ **Foundation.** This item will vary considerably based on factors such as slope of lot and the height of basement walls.

- ◇ **Rough Lumber.** A list on page 80 gives you an idea of what goes into the "framing package," as it is called. This includes all materials except windows, doors, and roof shingles (although it can include those as well). These are all the materials necessary to *dry-in* the house. A good lumber company will put together this material list for you free of charge because it wants your business.

- ◇ **Rough Labor.** This is the labor required to bring the house to the dry-in stage. After this stage is completed, all other stages can commence, some simultaneously. The best way to contract for this job is by the square foot, with the square footage agreed to before you start. Five people will arrive at five different square footage totals, using the same set of plans. Some will vary by 300 square feet or more. Sounds incredible, doesn't it? But I swear that it's true. In determining the square footage, houses are measured from outside wall to outside wall, *not* from roof overhangs. If the house is not easily divided into rectangles for simplifying square footage determination and you can't figure it out, have the designer do it for you. Ready-made plans generally come with the square footage broken down for you. Use those figures.

◇ **Windows and Exterior Doors.** This cost is simple to estimate since you have an exact count. I do not recommend any particular brand, but I do recommend that you visit a couple of building supply companies and compare. Most carry more than one brand. Locally made windows usually are less expensive than national brands and often carry comparable warranties. For special windows such as angular or bay windows, get exact quotes from the supplier. Generally there is no additional cost to install windows (except special or unusual ones), as that labor is included in your carpenter's framing charge. Be sure it is.

Exterior doors are as easy to estimate as windows, and all the same factors apply, including that the labor to install them should be included in the framing bill from your carpenter. Sliding glass doors will be slightly more expensive and may require a separate installation charge, depending on your carpenter. Be sure that the installer caulks under the threshold thoroughly, using exterior grade silicone or polyurethane caulk.

◇ **Roofing.** It is measured and estimated by squares. A square of roofing is the amount of roofing material required to cover 100 square feet. The number of squares can be determined by your supplier, your carpenter, or you. I advise having either of the first two do it for your first house. They won't be exact, but may come a little closer than you will. Asphalt-fiberglass shingles are priced by the guarantee offered by the manufacturer. Fifteen-year shingles will be less expensive per square than 20- or 25-year shingles. It is probably best to go with a better-quality shingle, which will still be more economical than other types of roofing such as cedar shakes or steel panels. The cost of labor to install will depend upon the sub, the weight of the shingle (more durable shingles are usually heavier), and the pitch of the roof — the steeper the pitch, the higher the price for labor. The same applies to weight. The heavier they are, the higher the price. The most common shingles for roofing are the 245-pound asphalt variety, and an average roof pitch is about 6 inches rise for each foot of horizontal travel (a 6/12 pitch).

◆ **Concrete flatwork (slabs), garage floors, basement floors.** This refers to smooth finish concrete work, not rough finish as in driveways, patios, and walks. It also involves the use of other materials such as Styrofoam, wire mesh, *expansion joints*, and *polyethylene*. Extra site preparation and gravel base installation can also figure in this expense. Your concrete subcontractors can explain this to you. The work is closely inspected by most building departments. Get a bid based on square footage of actual concrete area.

◆ **Siding.** Depending on the material used (which may range from brick to vinyl), you can get an accurate bid from a subcontractor that includes the cost of siding materials and any flashing required around windows and doors.

◆ **Plumbing.** This bid should include all fixtures such as toilets, sinks, and water heater. It will *not* include such items as a dishwasher, garbage disposal, washing machine, or other household appliances.

◆ **Heating.** Use heating and air-conditioning systems recommended by your local utility company experts. Cost of installation should include proper ventilation for bathrooms, the kitchen, and clothes dryer.

◇ **Electrical.** In addition to electrical wiring costs, this bid should include all switches, receptacles, wires, panels and *breakers*, wiring of all built-in appliances, cable and phone wiring, and compliance with codes.

◇ **Insulation.** Get a bid per local code for minimum insulation. To get the maximum insulation for number of dollars spent, consult your local utility company experts.

◇ **Water (Well).** For water tap-in fees call your local municipality. For a well bid, call a well drilling firm familiar with the area and get a firm maximum bid as well as a drilling price per linear foot drilled.

◇ **Sewer (Septic).** For a sewer tap-in fee, call your local municipality. For the price of a septic field, get a written bid from a local contractor who does septic system installation.

◇ **Fireplaces.** Whether it is masonry or prefab, get a bid in writing.

◇ **Drywall.** Bids should include labor and materials to hang the wallboard, tape joints, and finish with joint compound (two coats).

◇ **Cabinets.** Bids should include kitchen cabinets and bathroom vanity cabinets. Labor to install should be included in trim labor.

◇ **Interior Trim.** A bid from the lumber supplier should include all interior doors, moldings, closet shelves, and stairway trim. It should also include *carpet underlayment*.

◇ **Interior Trim Labor.** You should get a bid to install all the materials for both cabinets and interior trim, above.

◇ **Painting.** First-time contractors often skimp on this category, often planning to do it themselves. But it is still smart to get an estimate, even if just for the materials. If you are not planning on doing the labor yourself, get a written bid for both labor and materials.

◇ **Appliances.** In the planning stage, you just need to have an idea of what you want in the way of appliances and insert a dollar amount allowance in your estimate. At this stage you don't need the actual models picked out, just a ballpark idea of what you think you will eventually buy. Use a dollar allowance you feel is adequate to get the appliances that you want.

◇ **Light Fixtures.** Though you probably won't have your actual fixtures selected, figure an amount that will be enough to cover the costs of all necessary fixtures.

◇ **Floor Coverings.** Estimate approximate costs for all floor coverings, taking into account varying amounts for wood, carpet, tile, or other coverings.

◇ **Driveway.** Depending on the material used, get a bid in writing based on the square footage of the area to be covered. Make sure you agree on the square footage.

◇ **Garage Door.** If you can still afford one, get a price from your local lumber company — with or without operators, installed or uninstalled.

◇ **Other.** Anything not included above, i.e., fences, sidewalks, decks, swimming pools, saunas, etc.

Chapter 5

Further Preparations

Y‌OU HAVE PURCHASED a lot, selected house plans, estimated costs, and arranged your construction loan and permanent mortgage. Several details still need your attention before you break ground.

Obtain a building code book or pamphlet from your local building inspector, listed in the phone book. If you don't have a building inspection department, find out about any local or state codes and how they are enforced. This information can be obtained through your city or town hall, your lender, or an attorney. You need not be an expert on codes, but it is best to be familiar with a few important specifications, such as those that relate to the thickness of footings, the preparation of concrete slabs, and requirements for foundation walls and lumber spans. If you don't have a building inspection department, a local library should have a code book for your state and/or county, and several good reference books on each construction trade involved. You may want to do this before getting a loan. Regulations could change your plans — and increase your costs.

Again, I emphasize that you need not be an expert. Your subs will know — or should know — the codes as they apply to their particular fields. If there is a question, you'll know where to find the correct answer if you have a code book and reference books. If you find good subs, you may never need to use those references.

Also, at this point, you must find out from your building inspection department the procedure for obtaining your building permit. The sample form on the following two pages gives an example of information you will need and the inspections that will be required. Ask your local department about regulations in your area. Good subcontractors will take care of the inspections for you.

BUILDING PERMIT
DEPARTMENT OF BUILDINGS AND SAFETY ENGINEERING

The Bureau of Licenses and Permits
hereby grants permission to

Permit No. ————————————

Date ———————————— 20———

_____

name   address   contractor's state license

_____

to   ( ) story   number of apartments

_____

on the   side of (building no.) street or avenue   zoning district

_____

lot no. and subdivision   size

Building is to be ————— ft. wide by ————— ft. long by ————— ft. in height
and shall conform in construction to type

Use group————————————— Basement walls or foundation—————————

Remarks: _____

_____

Approved by_____

cubic feet————— estimated cost $————— regular fees $————— zoning fee $—————

Owner _____

Address _____

Bureau of Licenses and Permits

by _____

This permit conveys no right to occupy any street, alley, or sidewalk or any part thereof, either temporarily or permanently. Encroachments on public property, not specifically permitted under the building code, must be approved by the common council. Street or alley grades as well as depth and location of public sewers may be obtained from the Department of Public Works — City Engineer's Office. The issuance of this permit does not release the application from the conditions of any applicable subdivision restrictions. Minimum of three call inspections required for all construction work:

1. Foundations or footings.
2. Prior to covering structural members (ready to lath).
3. Final inspection for compliance prior to obtaining certificate of occupancy.

Approved plans must be retained at the job site and this card must be posted until final inspection has been made. Where a certificate of occupancy is required, such building shall not be occupied until final inspection has been made and certificate obtained.

Separate permits required for electrical and plumbing installations
POST THIS CARD

Building Inspection Approvals

1

Drain tile and foundation _____

Date _____

Inspector _____

2

Superstructure _____

(prior to lath and plaster)

Date _____

Inspector _____

3

Final inspection _____

Date _____

Inspector _____

Work shall not proceed until each bureau has approved the various stages of construction.

Electrical Inspection Approvals

1

Roughing in _____

Date _____

Inspector _____

2

Final inspection _____

Inspector _____

Safety Engineering Approval

3

Approved _____

Date _____

Inspector _____

Plumbing Inspection Approvals

1

Building Sewer _____

(A) Sanitary _____

Date _____

Inspector _____

(B) Storm _____

Date _____

Inspector _____

2

Crock to iron _____

Date _____

Inspector _____

3

Underground storm drains_____

Date_____

Inspector _____

4

Rough plumbing _____

Date _____

Inspector _____

5

Water piping final inspection_____

Date _____

Inspector _____

**Inspections indicated on this card can be arranged for
by telephone or written notification.**

Inspections Required

As an example, the county I live in requires these inspections.

◇ Temporary electrical, or *saw service*, to ensure proper grounding.

◇ Footing, before pouring concrete, to make sure you have reached solid load-bearing ground.

◇ Slab, before concrete is poured, to determine whether it is properly insulated. Any plumbing in the slab is also inspected.

◇ Electrical, plumbing, heating, and air-conditioning *rough-in*, to ensure that in-the-wall installations that can't be seen later are correct.

◇ Framing rough-in, to ensure structural integrity, especially after electrical, plumbing, heating, and air-conditioning installations. Workers have been known to cut too deeply into a joist or stud and weaken it.

◇ Insulation, to ensure compliance with new local standards.

◇ Final electrical, plumbing, heating, and air-conditioning and building to ensure they work properly, comply with codes, and are safe. All final inspections must be completed before residents can get more than temporary electrical service.

If you haven't already done so, now is the time to select materials such as brick, shingles, windows, doors, roofing, trim, plumbing, fixtures, built-in appliances, light fixtures, flooring, hardware, and wallpaper. As you visit building supply houses to see their samples, you can also take care of a couple of other important tasks:

◇ Open your line of credit with them. This is quite easy, even for an inexperienced builder. See Chapter 7 for a list of the types of suppliers you'll need.

◇ Ask them to recommend reliable subcontractors. Smaller building supply companies are better equipped to furnish this information because the management is more directly involved with customers. The key sub you are looking for is your carpenter, and most of the good ones patronize building supply houses from time to time. Many make building supply houses their source of referrals. This is especially true in resort areas. (Don't forget, you can also use this book as a guide to building a second home in a resort area.) You may find other subs here. More in the next chapter about subs.

Now is the time to contact your gas, electric, and water departments for hookup procedures. Every locale has different procedures. You may also want to order a *prewire* from your phone company at this time, as the company sometimes requires three to five weeks' notice. Some don't charge for this, some do. Ask first.

You will also want to shop around for a builders' risk or fire policy from various insurance agents. In most states, agents are fairly equal in policy offerings, since they're governed by state laws. But be sure to check. The policy should take effect when materials arrive on the job, but your lender may want it to be in force before closing the construction loan. The lender will be the payee in the policy since it is the company's money. When the house is complete, and you move in, the fire policy can be converted to homeowner's insurance, often at a fairly good savings over a new homeowner's policy.

Doing Your Own Labor

If you plan to do any of your own labor, I have only one thing to say: If you aren't an expert at the particular trade you plan to do yourself, forget it. This is especially true in painting. People think it is so easy, but unfortunately it comes at a time when the house is quite far along, and construction disbursements will be approaching their maximum amount. You need painters who can get in and get out, do the job fast, and at the same time do a respectable job. If you seek perfection in painting, wait until you are living in the house and do your own touch-up painting.

If you hire painters by the hour, plan to stay with them the whole time, or plan on it taking twice as long. You won't get a better job for the extra time. Hire them by a contract amount, such as ninety cents a square foot based on the amount of heated area.

When it comes to doing your own work, remember one of Murphy's laws — "Nothing is as easy as it looks."

Chapter 6

Subcontractors

A SUBCONTRACTOR (sub) is an individual or a firm that contracts to perform part or all of another contract. In your case you are technically the builder or general contractor, and you will build by subcontracting with others. You will pay for this by a predetermined contract amount with each one. This is important. You will have no hourly wage employees working for you. Thus you will avoid all of the governmental red tape and taxes concerning employees. Your subcontractors are not considered to be employees.

Your Subcontractors and Professionals

Here is a list of the subcontractors and professional people you probably will be contracting with, listed generally in the order in which you will need them.

◆ Attorney. Needed for initial assistance for questions, land purchase, loan assistance, loan *closing*, continuing assistance if necessary for legal questions and/or settling disputes with subs or suppliers. I would not attempt the project without an attorney. You can find one who specializes in real estate by calling the Lawyer's Referral Service (in the Yellow Pages) or a real estate company, or asking a friend. Attorney's fees are a cost of construction.

◆ Lending officers at banks or savings and loans, or mortgage companies.

◆ House designer or architect.

◇ Framing carpenter (your key sub, to be lined up early).

◇ Surveyor.

◇ Grading and excavation contractor.

◇ Footing contractor.

◇ Brick masonry contractor. Builds the foundation.

◇ Concrete (finisher) subcontractor. Pours the slab or concrete floor.

◇ Waterproofing contractor.

◇ Electrical contractor.

◇ Plumbing contractor (and well and septic system, if needed).

◇ Heating, ventilation, and air-conditioning (HVAC) contractor.

◇ Roofing contractor.

◇ Insulation contractor.

◇ Drywall contractor.

◇ Painting contractor.

◇ Finish carpenter. Installs kitchen and other built-in cabinets and trim around doors and windows. This work is often, but not always, done by main carpentry sub.

◇ Flooring, carpet, and Formica contractor.

◇ Tile contractor.

◇ Cleaning crew contractor.

◇ Landscape contractor.

Finding Your Subs

As mentioned earlier, a lumber supply store is the best place to start for finding your carpenter subcontractor. Your carpentry sub will be able to recommend almost everyone else, as he is on the job more

Carpentry Labor Contract

To: (your name)_____ Subcontractor_____
(address)_____
_____

Date: _____ Job address _____

Owner: _____
Area: _____ Heated _____ sq. ft.
                                      Unheated _____sq. ft.
                                        Decks_____ sq. ft.

Charges

Framing @_____ sq. ft × _____ sq. ft. = $ _____
Boxing and siding @_____ sq. ft × _____ sq. ft. = $ _____
Interior trim @_____ sq. ft × _____ sq. ft. = $ _____
Decks @_____ sq. ft × _____ sq. ft. = $ _____
Setting fireplace $ _____
Setting cabinets $ _____
Paneling $ _____
Misc. $ _____
Total charges $ _____
Signed: (your name) _____ Date _____
Signed: (subcontractor) _____ Date _____

than anyone else and knows most of the other subs involved in building a house.

"A good sub is a working sub," especially during a recession or other downturn in housing starts. This is not always true, but it is a pretty safe bet. The really good ones are sought after and always busy because they do good work and are reliable.

If you can't find a sub through a supplier or your carpenter, the next best place to look is on a job. Find a house under construction. Stop and ask around. You can get names, prices, and references. This takes only a few minutes. It is done all the time and the general contractor shouldn't mind. He probably won't even be there. Besides, he does (or his superintendents do) the very same thing.

Often the boss or owner of a subcontracting firm is on the job. Get his number and arrange a meeting. Sometimes there are signs at the job site advertising different subs.

Only certain trades of subs are listed in the Yellow Pages. Most independent carpenters and brick masons are not listed. (Many of the listings under "Carpenters" or "Mason Contractors" may be for general building contractors.) You'll probably find heating and air-conditioning companies, plumbers, electricians, roofers, waterproofing companies, lumber dealers, and appliance manufacturers.

I had someone who was probably one of the best drywall subcontractors in the Southeast, but he was not listed. He didn't live in the same city; he lived out in the country. He was so good that he stayed backlogged two or three weeks even during recessions. His name was given to me by a lumber dealer.

Subcontractor's Contract

Subcontractor: _____

Address: _____

Builder: _____

Date: _____ Plan No. _____

W. Comp. Ins. Co. & Agent _____

Certificate No.: _____ Expiration Date: _____

Location of Work: _____

Total Price per House: _____($) Dollars

Terms of Payment: _____

Work to Be Performed and Materials to Be Supplied:

_____

_____

_____

_____

_____

For Subcontractor: _____
                         signature and title

For Builder: _____
                         signature and title

Each subcontractor should carry insurance on his or her employees and should provide you with a certificate of insurance. (See the example in the Appendix on page 117.)

It should not be a problem if most subs in your area belong to unions. I state this because some people expect union labor to cost more or fear they will have to do excessive paperwork if they hire union members. Not so. Your subcontractor will be responsible for all compliances to the union, and the prices should be competitive.

Since this is your first experience and you won't be familiar with prices in your area, get three or four bids, or quotes, before selecting a sub. Use a written contract with all subs.

You may use the very simple ones I have provided or your attorney can provide you with one. Subcontractors may have their own contracts. At any rate, use one. Don't trust anyone's memory when it comes to dollars. Be sure your bids are comparable, that all are bidding on exactly the same work. If your specifications are properly spelled out in your plans, the bidding should be comparable.

Bids for Plumbing

Plumbing bids should include all plumbing fixtures right down to the toilet seat. They will not include accessories such as toilet paper holders. If colored fixtures are to be used, specify color and brand. Plumbing showrooms are your best bet for the selection of these fixtures. Magazines and brochures don't tell you enough and don't often give prices. Most plumbing showrooms won't tell you the wholesale price, but you'll be paying list anyway, as the plumber makes a profit on each fixture and it's included in the bid. Don't make an issue of this. The small profit in the fixtures is one of the plumber's sources of income and he earns it.

Your heat and air-conditioning contract should include vents (generally fan powered) for the bath, clothes dryer, stove, and range hood. Electrical bids should include all switches, wiring, receptacles, circuit breakers and their respective *panel boxes*, a temporary service box and installation, saw service, wiring of all built-in appliances, and installation of ovens and ranges, furnaces, heaters, and air conditioners.

All subcontractors should be responsible for obtaining inspections from the building department, but you will want to be sure they do or you will have to do it yourself. Lack of inspections can cause delays. Proceeding without getting inspections can be troublesome and expensive. If you, for example, put up drywall before having your wiring inspected, you could be made to tear down some of the drywall for the electrical inspector. This is not likely to happen, but the inspector could force the issue.

Utilities must be connected. Exactly who is responsible for running water lines, sewer lines, and electrical hookups will vary with each subcontractor involved. Get the responsibility pinned down when you are hiring the subs, then follow through to be sure it is done properly.

Scheduling Your Subcontractors

Try to schedule your subs to fit into the sequence of events outlined in Chapter 8. This won't always be possible, and one sub can hold up the process. This is why you should check their references and ask whether their previous work has been completed on time. Reliability is as important as quality and in some cases more so.

Paying Your Subcontractors

When you sign your contract with your carpenter, you will agree on a contract price for the work. It is usually based on X number of dollars per square foot of heated area and X number of dollars per square foot of under roof, such as in the garage. Prices will vary with the area, unions, and the complexity of the job.

Never pay a sub for work not done, for work that is incomplete, or for an unsatisfactory job. Never pay in advance.

Work out a schedule of payment with your carpenter. The carpenter and some of the other subs may require draws, or partial payments, as work progresses. This should be discussed before work begins. Don't be shy about it. They are accustomed to discussing such matters.

It is all right to pay a draw, but never pay more than for work done. If, for example, your carpenter says he is 50 percent finished framing, but he has only completed the floor framing and wall framing and hasn't finished the ceiling joists, roof framing, sheathing, or *bridging*, he isn't 50 percent complete. He is nearer 40 percent complete.

If he is charging $1 per square foot of area under roof (including the garage, storage, and porches) you will owe him $2,400 total when he has finished framing a 2,400-square-foot house. At 40 percent complete, you would give him a maximum of 33 percent, or $800.

Plumbers and electricians usually get 60 percent of the total contract price when their rough-in has been completed and inspected. Heating, ventilating, and air-conditioning rough-in payments depend on the installation of equipment such as furnaces. If payment is just for duct work and some low-voltage wiring, 20 percent of the total should suffice. If a furnace had to be installed during rough-in, add another 10 percent. Work out the arrangement with the sub before he starts. Subs almost always would like to get more money than they have in the job. Be sure there is enough money left in the total bid to complete the job if the HVAC dealer goes broke while you are still building. It has happened. You don't want to be stuck paying more to complete his job. You'll be covered better if you don't overpay him on his rough-in.

Brick masons and painters are about the only other subs who will require a draw in progress. You will have to use your best judgment as to how much of the job is done. Again, don't get ahead of them in paying.

I seem to be saying the only way you'll get your sub to complete the job is if you owe him money. In some cases that is true, but in others it is only partially true. Some subs would finish regardless. The main objective is to get your job finished. Often subs will have more than one job going at one time. You want yours finished quickly and done before one that was started after yours.

Make sure inspections by your county or city are completed and the work is approved before you make any payments at any phase of construction, other than partial draws. This is your assurance that the job will be done, and done properly.

Working with Subcontractors

If you get along well with everyone at all times, you may not need to read the next few paragraphs. But if you occasionally run into conflicts, read them carefully. The fault — sometimes — may be yours.

At this point you've selected your subs. You're satisfied, now that you've checked them out, that they are honest, trustworthy, and experts in their fields.

Now let them work. Don't try to supervise every blow of a hammer, the placement of every stud. They know more about their trades than you do, and probably, if they came to you well recommended, they take pride in their work. Let them do it.

And, more emphatically, don't try to tell your subs their jobs just because you have read this book and a few others. You'll get good work out of your subs if they understand you think they know their jobs, and you're depending on them for good advice and quality work.

When a sub's work is completed, when the work looks good, and when the related inspections have checked out, make sure to promptly pay the contracted amount. A hearty thank you is also in order. Subs who get treated right throughout the job and afterward will do a better job for you, and they'll come back when you build your next house.

Chapter 7

Suppliers

THE SUPPLIERS THAT you will be buying from are listed in this chapter by the type of products they sell. Some sell several different products, so you can shorten your shopping time when you patronize them. All of them will require a credit check on you by means of the references that you give them. Usually three charge accounts (such as Sears) and one bank reference are required. Your suppliers will also want the name of your lender. This lender reference is the key to your getting credit with them. Obviously, if your credit is strong enough for the lender, it's strong enough for the suppliers. Often, just the lender reference is enough.

The suppliers you most likely will be using are:

◆ Sand and gravel company.

◆ Brick company. For face (decorative) brick.

◆ Concrete block and brick company. For mortar mix and the building blocks for foundations.

◆ Concrete supply company. For concrete for basements, garages, footings, and driveways.

◆ Lumber company. For framing lumber, nails, windows, doors, roofing, siding, paneling, and interior and exterior trim.

◆ Flooring company. For vinyl, wood, and tile flooring and for carpet.

◆ Cabinet shop. To fabricate plastic laminate countertops.

◇ Lighting fixtures supplier. For light fixtures, bathroom fans, exhaust fans. The firm will do a complete count of your lighting needs and help you stay within a dollar figure set during estimation.

◇ Paint store. For wallpaper. Paint is usually purchased by your painter and included in the price per square foot.

◇ Appliance store. Some lumber companies and lighting centers carry appliances.

◇ Insulation company.

◇ Tile company. For ceramic tile, marble, and any decorative stonework.

◇ Drywall company. For Sheetrock. Some lumber companies also sell Sheetrock.

◇ Any other specialty type suppliers that you may need for certain items not carried by one of the above, such as cabinet shops. Most lumber companies sell cabinets.

An account at a plumbing supply company is not necessary because the plumber sub buys all these items. You may need to visit to make your selections.

Some of the suppliers listed may carry most of your needs. In some small towns I've seen lumberyards that carry everything from bricks to wallpaper. So long as their prices are competitive, that's fine.

This is my recommendation: When you find a lumber company, give your plans to the sales force and let them give you a list of all the items the company can provide and a complete cost list. I have never found a company that wouldn't do this. Some will do a complete take-off of all the lumber and other materials you will need.

You will get a list of the number of studs, floor joists, windows, and rafters. It will be fairly accurate, so don't be afraid to use it. Remember, the company is eager to sell to you and this is part of its service. As the housing market gets slower, the company will become more eager. Company personnel can also explain different products to you and show you new products and ideas. Let all your suppliers aid you in take-offs and different product ideas.

Delivery of Materials

Most companies will also assist you in, or be responsible for, delivery, making certain the materials are delivered on time, but not days in advance so they may be stolen. Make sure there's a flat, clear area close to the foundation where the load of framing lumber can be off-loaded. This delivery system has worked particularly well for me when I have used one supplier for all framing and as much of the rest of the supplies as I could. In such a case I have permitted my carpenter to order needed materials from that supplier alone. The supplier has aided me in keeping an eye on the additional materials ordered.

Buying at Builders' Cost

"Ask and ye shall receive." If you don't, you won't. Be sure to ask whether the quoted price is the "builders' price." Tell the supplier that you are a builder, because you are. If you want to add "company" to your name (that is, John Smith Company), fine. If it makes you feel better, do it. It costs nothing and changes nothing legally (the credit will still be in your name). If you want to incorporate (form a corporation), talk to your attorney. I don't think you'll find it is necessary.

Paying Your Suppliers

Some suppliers offer at least 30-day terms so that you pay the stated amount within that time without penalty. They may also offer 2 percent discounts if your bill is paid by the tenth of the month following purchase. If you buy supplies on June 20, you must pay by July 10 to receive the discount. If you buy on June 1, you would still have until July 10 to pay and get the discount. I try not to order too much toward the end of the month and wait as much as a week, if I can, to get an extra 30 days to pay. Not all suppliers are as generous as this, and their terms will vary. Check with your suppliers on what terms will be offered you.

By getting favorable terms and extra time in which to pay, you can be sure that your construction draw will cover the bill. You will use construction draws to pay for all your supplies. You should never have to use any of your own money.

Construction draws are based on labor completed and materials in place, not stored or sitting on the job site. For example, if you order brick for the whole house (a *brick veneer* home) at the start of the job, you'll have to pay for the brick that's going to be used in the veneering before the lender advances the dollars for veneering. Instead, you should order the brick in two stages, first for the foundation, then for veneer when you're ready.

Bookkeeping

Bookkeeping is very simple when you are building only one house. Open a separate checking account to handle the construction of the house and pay all bills, no matter how small, by check. Code each check stub to one of the appropriate categories on the estimate sheet and record each amount on a sheet of paper until you have totaled all of the money spent on that category. Then enter the total amount in the actual cost column of the estimate sheet and compare the actual cost to the estimated cost. You may find the use of a good computer program is helpful in your bookkeeping efforts.

Chapter 8
Building the House

LOOK AT HOW MUCH you have accomplished. You have your building lot, you've arranged your loan, and you have a house plan you can live with and are able to finance. You've completed the vast amount of paperwork that includes any necessary permits and insurance policies.

You have located most of your subs and have contracts with them. You've visited various supply houses and worked out your accounts for bricks, concrete, and lumber.

Congratulations. You've reached the day you thought might never come. You're ready to start building.

What is the proper sequence of steps in building the house, and how long will each take? Let's make a list of them.

1. Staking the lot and house: 1–3 hours.
2. Clearing and excavation: 1–3 days.
3. Ordering utilities, temporary electric service, and a portable toilet: 1 hour.
4. Footings (steps 3 and 4 can be reversed). First inspection must be made *before* pouring: 1 day.
5. Foundation and soil treatment, then foundation survey: 1 week.
6. Rough-ins for plumbing, if on a slab, and inspection: 2–4 days.
7. Slabs, basement, and garage: 1–2 days.
8. Framing and drying-in: 1–3 weeks.
9. Exterior siding, trim, veneers: 1–3 weeks.
10. Chimneys and roofing: 2 days–1 week.

11. Rough-ins, can be done while steps 9 and 10 are in progress: 1–2 weeks.
12. Insulation: 3 days.
13. Hardwood flooring and underlayment: 3 days–1 week.
14. Drywall: 2 weeks.
15. Priming walls and "pointing up": 2 days.
16. Interior trim and cabinets: 1–2 weeks.
17. Painting: 2–3 weeks.
18. Other trims, such as Formica, ceramic tile, vinyl floors: 1 day–1 week.
19. Trimming out and finishing plumbing, mechanical, and electrical and hooking up utilities: 1–2 weeks.
20. Cleanup: 2–3 days.
21. Carpet and/or hardwood floor finish: 3 days–1 week.
22. Driveway (if concrete, can be poured anytime after step 14): 1–3 days.
23. Landscaping: 1–3 days.
24. Final inspections, surveys, and closing of construction loan and interim loan: 1–3 days.
25. Enjoy: a lifetime.

(Note that steps 2 and 4 can be done by one sub.)

The Steps Explained

Let's take a closer look at those steps and clear up any details you should know about them.

STEP 1: Staking the Lot and House (1–3 hours)

The number one problem that occurs when building is the placement of the house on the lot. Staking is vitally important in the early stages of the building process. Since this will be, most likely, the first house you have built, I recommend that you have a registered surveyor/engineer stake both the lot and the position of the house. The least expensive one should be the one who surveyed the lot for purchase. Surveying is an important function, as houses have been placed in violation of certain setbacks or restrictions, and if a surveyor does this, he or she is responsible. Builders have placed houses straddling property lines and have had to tear the foundations out and start over. Best to be safe.

The stakes for the lot may have been moved or torn down since you purchased it. A surveyor will check this. Cost for restaking a lot and staking a house should be $100 to $200, depending on the complexity of the house and the terrain. You, of course, will want to be present for the staking or placing of the house to be certain it faces the direction that you want. If possible, meet with the surveyor at the site before he or she begins working. Or you could

You can be a successful contractor of your own house without knowing a soffit from a fascia, but it helps, in dealing with the subcontractors you hire, if you can speak their language. You'll find in this illustration a sample of the common terms you'll hear used by your carpenters. For other terms of the trade, and some you need to know when dealing with your home financing, see the Glossary on page 90.

put some stakes in the ground over a weekend and thus indicate the approximate location and direction you desire. Then let the surveyor do it accurately and you can inspect it when it is done.

When the lot is cleared and the basement — if you have one — is dug, you may want the surveyor back to restake the house. On his first visit he usually will put in offset stakes. Offset stakes are the original four or two corners offset as much as 25 feet to the side so they won't be disturbed during clearing and excavation. Then it's a quick job for him to find the exact locations on his second visit.

Usually your house should face in the same direction as other houses on your side of the street. It doesn't have to, especially if your house is on a large lot, unless the building code demands this.

If your lot slopes more than 3 or 4 feet, you may need a *topographical plat*. Shop for the best price. You need it so that you can fit the house to the slope and be certain water will go around the house as it drains off the grade.

Water is, has been, and always will be the biggest problem that interferes with the construction of a building. It is nature's strongest force. It will go where it can, and it will take the shortest route. It is stronger than anything man can make. Gutters help get rid of water; they don't control it. Keep that in mind.

The house should be positioned first on a map of your lot, then that position should be staked on the site by you and your carpenter, surveyor, or footing contractor. The surveyor is best for the job. An architect or designer can position the house on the lot on paper, considering water as well as solar energy, wind, light, and the location of other houses. You should contribute your ideas to the person positioning the house. A few things to consider:

- ◇ Light. A north-south facing house will be darker than an east-west facing house.

- ◇ Water flow.

- ◇ Other houses on the street.

- ◇ The street it faces. Should the house be parallel with the street? What if the street curves? What about a corner lot?

- ◇ Privacy, on the front, back, and sides.

◆ Solar orientation. To get the most benefit from the sun's energy, homes should have as much surface as possible facing south, with broad expanses of glass on that side and a minimum of doors and windows on the north side. Homes may also be designed to allow for roof solar collectors or other solar energy devices.

◆ Minimum setback and side-yard requirements.

I don't want to frighten you with so many considerations. I have built many houses for my family, and I still have to think through these decisions. Positioning the house is a very personal decision, one you should make when you buy a lot. If you are stumped and you can't decide for yourself, and you don't have an architect or designer involved with your plans, spend the money and get the advice of an architect for that particular decision.

STEP 2: Clearing and Excavation (1–3 days)

Clearing the lot is what the title implies, clearing trees, brush, rocks, roots, and debris from where the house will sit, and usually 10 feet or more around the foundation, thus allowing space for tractors, fork lifts, and trucks working at the site. Obviously, the more area to be cleared, the more it will cost. Big tree removal is time-consuming and expensive. Rocks may have to be blasted. If you want unwanted trees to be cut into firewood, the crew you hire will charge dearly for doing the job. Best to have the good, manageable-sized logs cut down to 10 to 12 feet in length and piled at the side of the lot for you to cut up at your leisure.

Your best source for finding a sub for clearing and excavating is by word of mouth or in the Yellow Pages.

Get a *contract price* for this work. It may cost a little more, but you will be assured of not having your first cost overrun. If a basement is to be dug, your sub must have and know how to use a *transit*. You may also want your surveyor/engineer to oversee this digging to make certain it is the proper depth. Some or all of the dirt removed from the basement site might be put out of the way for later backfilling and landscaping. Topsoil should be separated, to be spread later for the lawn or garden.

Your contract price should include hauling all trash, such as stumps, branches, and rocks, to a suitable landfill. I don't advise burying trash on your lot, as these stump holes tend to form an unsightly depression as the material in them settles or rots. Some areas ban stump holes. Of course, if the distance to a suitable land-fill is prohibitively expensive and you have enough land, a stump hole may be your best choice.

You may want to put one or more loads of stone on your drive-way so that supply trucks can drive in during wet weather. The best stone for this is unwashed crushed stone. It has all the powdery sub-stance created in crushing the stone, and it will harden after getting wet. You also may need to put drainpipes in at the roadside if they are required by either ordinance or common sense. They allow road-side water to flow under the driveway and prevent water from wash-ing the stone away.

STEP 3: Utilities Hookup (1 hour)

When you purchased your lot, you were told — be sure you were — what utilities were available and how much they would cost. Now it's time to make plans for a couple of months down the road with a few phone calls and/or a visit to each utility. Pay all fees and complete any necessary forms. Arrange for temporary electric service for your subs. Usually your electrician is responsible for installing the temporary electrical panel box and having it inspected, but *you* will have to apply for the service from the utility. This usually can be done over the phone.

Wells and *septic systems*, if used, can be installed now, and it is best to get this work done at this time. County or city health inspec-tors may be required by code to determine the location of these. Tell them your plans for such things as gardens or driveways, or which trees you hope to save, to guide them in their decisions.

If no temporary source of water is available, such as a house next door, you will have to have the well dug and temporarily wired for your brick masons, who will be needed shortly, or they will have to truck in their own water.

I also recommend, and some locales require, a portable toilet on the job site. Sources for renting these can be found in the Yellow Pages under Toilets-Portable.

STEP 4: Footings (1 day)

The footing is the base of a structure. It is a mass of concrete supporting the foundation of the house. It can be poured into wooden forms or in trenches. It must be below the *frost line*, or it will heave when the ground thaws and freezes. In the northern states and higher elevations of any area, this line may be 4 or more feet below grade level. This is one reason there are more basements in a northern climate. If you have to be several feet below grade level for your footing, and thus need several feet of foundation to get back up to grade level, only excavation and a concrete slab are needed for a basement. Local codes will clearly state the requirements for footings in your area. The subcontractor you choose should know the code.

I have a footing sub who stakes, clears, excavates, digs, and pours footings. For your first house, I recommend that you find one who does the same. The cost will be comparable. The footing is probably the most important part of the house. If it settles or moves, so will your house. If it is not done according to the dimensions of your plans, you will have to change the plans to accommodate the footing or do the footing over. I recommend the former if the situation arises, unless the deviation is too severe.

After your foundation walls are up, put in a footing drain. Your code may require this. The drain can be connected to a dry well, storm sewer, or any other approved means of getting rid of the water. In some places, it can simply drain into your yard.

As a rule, footings are better today than they were 100 years ago. Well-built houses of today will probably last years longer than those built long ago. Technology has improved materials such as concrete, and our knowledge of how to use them. I say this to help ease your mind about this important step.

Building inspectors usually check the locations of footings before they are poured to make certain they are deep enough and resting on undisturbed earth. Don't complain about this inspection — it could save you thousands of dollars if it means that you will avoid some future problem.

STEP 5: Foundations (1 week)

Foundations can be made of brick, concrete block, or poured concrete. Stone foundations, as a rule, aren't built anymore, as they

•

aren't as strong as the others. Stone can be applied as a veneer just like brick, for aesthetic purposes, and you would be wise to use it only as such. Local codes may prohibit stone as a foundation load-bearing material.

Your masonry contractor needs to be one of your better subs. Next to your carpenter, he is the most important. Your carpenter probably can recommend a good mason, as he usually starts his work soon after, if not directly after, the mason is finished. He knows the good ones. He's had to, I'm sure, follow bad ones. If he followed a bad one, he'll remember having to *shim* walls or compensate for an out-of-square foundation. Houses *can* and should have square corners.

The foundation wall for any type house needs to be high enough so that water (our old friend) will be diverted away from the house by the final grade of the soil around the house. It must also be high enough so that the wood finish and framing of the house will be at least 8 inches above the finish grade and thus protected from soil moisture.

A crawl space should be at least 18 inches high so that you can crawl beneath the house annually to inspect for such things as termite damage. The crawl space walls should have screened openings for ventilation.

If you are planning on a full basement, your foundation walls should be high enough so that you have at least 7 feet 4 inches of headspace in the finished basement.

How to establish the depth of excavation.

If you are in doubt about the foundation height, consult an engineer. Usually you, with the help of the carpenter, mason, excavator, or anyone who can use a transit or a level, can determine the needed height. If the lot is almost flat, the job is simple. It becomes tricky when the lot is steep or has opposing grades. Experienced contractors make certain the foundation is high enough at the highest point of the outline of the foundation wall, and use that highest point as the control point.

The finished foundation should be waterproofed from the footing to the finish gradeline. I recommend hiring a professional waterproofing company for this. Companies are listed in the Yellow Pages under Waterproofing. Don't let a laborer do it in his spare time, if he offers to. A professional company will stand behind its work.

Also, depending on your locale, you may need to have the soil treated for insects and pests, particularly termites. Hire a professional. This job is done after the foundation is in, but before any concrete is poured for either the basement or the garage. The cost is small.

It may come as a surprise to many first-time home builders that the foundation is poured or formed with concrete blocks, then holes are punched in it for such things as the water supply and the sewage outlet, the pipe is placed through the hole, and the space between the pipe and wall is patched. This is the easiest method for getting a tight, waterproof fit.

STEP 6: Rough-in Plumbing (2–4 days)

If you have a basement with plumbing or if you are building the house on a concrete slab (as opposed to wood floor joists), once the foundation is in and backfilled and *tamped* (packed down), and the soil treated, your plumber needs to install the sewer line and the water pipes that will be under the concrete. Also, any wiring that will have to be under the concrete needs to be placed in a conduit and roughed-in. Most wiring, though, can be run through the stud walls and ceiling joists to any given point.

Your treatment company may want to wait until the rough-ins are completed before treating the soil so it won't be disturbed by digging in the plumbing lines. Ask about your company's policy.

The plumbing system of a home. Note that the supply system pipes are black, the drainage system's are white.

STEP 7: Slabs — for Heated Areas (1–2 days)

Many locales require slab perimeter insulation. The most common perimeter insulation is extruded polystyrene foamboard, which is available in various thicknesses. The board extends from the top of the foundation to at least 12 inches below finished grade. I recommend a 1-inch or thicker board, even if it is not required by codes. I also recommend using a plastic barrier of 4 to 6 mil thickness under the concrete to prevent moisture from working up into the concrete. A 6-by-6-inch #10 wire mesh should be placed in the concrete to reinforce it.

The top of the slab should be at least 8 inches above the finish grade. Your sub should put down a base for the slab, tamping down gravel or crushed stone to form a layer 4 to 6 inches in depth. The poly goes down on this just prior to pouring the concrete. If you cannot cover the entire area with one sheet of poly, any joints of the poly should overlap by 4 inches and be sealed with silicone caulk. The wire mesh is laid on top of the poly. Call for an inspection before pouring concrete if your code requires it.

A good concrete sub will do all of this. I stopped checking my slab pourings when I found a terrific concrete subcontractor.

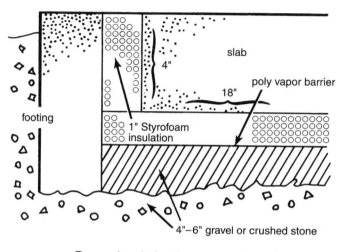

Proper insulation for a concrete slab.

For garage slabs, with proper backfilling and tamping, you don't need the wire mesh, but if you believe the added strength is needed, use it. I recommend an expansion joint of fiberboard around the perimeter. Garages are subject to extreme temperature changes, and concrete expands and contracts with those changes. The expansion joint permits this expansion without cracking the concrete. Both this slab and the house slab should be at least 4 inches thick. Be sure your sub thickens all slabs wherever they will be carrying load-bearing posts or walls. Codes differ on the additional thickness required, but it is often the thickness of the footings.

STEP 8: Framing (1–3 weeks)

If you have a good carpenter, this is the end of this section of framing. And if that sounds too simple, wait and see. You need only order the lumber, the windows, and the exterior doors, and in two or three weeks you'll have a house (or at least something that looks like a house). I have included here a typical order list for framing and drying-in, which means making the house secure from rain. Rain or snow during framing is not desirable, but it seldom does much damage beyond an occasional warped piece of lumber. But once the house is dried-in, the work inside can progress regardless of the weather if windows and doors are in place.

FRAMING LIST

15	pieces	2 x 6 x 12 treated pine
10	pieces	2 x 4 x 12 treated pine
20	lineal feet	2 x 2
250	lineal feet	1 x 4
18	pieces	2 x 8 x 12
80	pieces	2 x 8 x 14
80	pieces	2 x 8 x 16
55	pieces	2 x 10 x 12
100	pieces	2 x 10 x 14
18	pieces	2 x 10 x 16
18	pieces	2 x 10 x 18
2	pieces	2 x 10 x 20
14	pieces	2 x 6 x 14
45	pieces	2 x 6 x 16
80	pieces	2 x 6 x 20
800	pieces	Studs, 2 x 4 x 93
150	pieces	2 x 4 x 14
50	pieces	2 x 4 x 10
175	pieces	½-inch CDX plywood
24	quarts	Construction adhesive
10	rolls	#15 felt
60	pieces	Asphalt impregnated sheathing, ½ x 4 x 8
300	pounds	16d nails, coated
150	pounds	8d nails, coated
10	pounds	8d steel cut masonry
25	pounds	⅞-inch galvanized roofing nails
50	pounds	1½-inch galvanized roofing nails
10	pounds	16d galvanized finish nails

You will want to check with your carpenter about any problems with materials. It is at this stage that I recommend installing a job-site telephone, particularly if you are unable to communicate easily through cell phones, pagers, or frequent visits to the job site. Your carpenter can take the phone home at night to prevent misuse. Because it is impossible to estimate exact needs in materials, some ordering will have to be done during the course of construction.

double hung   horizontal sliding   fixed unit and awning

basement awning   casement

hopper

Various types of windows.

You could give your carpenter permission to order what he needs and then tell the supply house what you have done. The decision is up to you. I do it.

The drawings of typical framing shown below are to make you familiar with some of the construction terms.

When the framing is completed, order cabinets, bookcases, and vanity cabinets, if there are any. Space for them can now be measured on the job by the salesman.

Exterior wall framing at
the gable end.

Diagonal bridging of floor joists.

Framing around exterior wall
opening using header and
double studs.

A wood frame wall with masonry veneer. Note shelf on foreground of foundation, to support bricks.

STEP 9: Exterior Siding, Trim, Veneer (1–3 weeks)

This phase of construction is carried on while work progresses inside and should be done before roof shingles are installed. Masonry chimneys are installed after siding or veneer. Veneers such as brick should be installed before final exterior trim (boxing) is added. At completion of this step, you are ready for exterior painting.

STEP 10: Chimneys and Roofing (2 days–1 week)

Chimneys should be built before the roof is shingled. This will allow placement of sheet metal flashing around the chimney for waterproofing, and will also avoid damage to the shingles. A *prefab fireplace* and flue would also be installed at this time. Roofing follows completion of the chimneys.

STEP 11: Rough-ins (1–2 weeks)

All electrical, plumbing, heating and air-conditioning, phone prewires, cable or satellite TV lines, stereo and intercom, and burglar alarm systems should be roughed-in at this time or anytime after step 8 is completed. This does not mean that these units are installed at this time — only the wiring or plumbing for them. Inspections are needed when this step is complete.

STEP 12: Insulation (3 days)

Consult with your local utility company on the insulation you need to qualify for their lowest rates. Some locales require an inspection of insulation by both the utility and the building inspection department when it is completed, and before it is covered with drywall or paneling (or plaster, if anyone still wants plaster walls).

STEP 13: Hardwood Flooring and Underlayment (3 days–1 week)

I find it easier and neater to install hardwood flooring and vinyl or carpet underlayment before I have the drywall installed. It can be done afterward.

STEP 14: Drywall (2 weeks)

Most residential walls are drywalled with ½"x4'x8' gypsum wallboard sheets called drywall or Sheetrock. For baths and other moist areas use a waterproof board or paint it with an enamel paint, even before wallpapering.

Figure on 3½ or 4 times the square footage of floor area for the total square footage of wallboard to be used. Your drywall sub can give you a price based on a square foot charge. Some will give a bid that includes materials. On your first house you may want to go this route.

STEP 15: Priming Walls and "Pointing Up" (2 days)

After the drywall is finished and before the interior trim is applied, I prime all walls and ceilings with a flat white latex primer. My two painters do this with a spray gun and in one day can do a house that measures 3,000 square feet. By priming first, the finish painting time is reduced considerably, thus saving you money. With no trim installed, only windows need to be protected during spraying. If all of your woodwork (trim) is going to be painted instead of stained, priming the walls by spraying can be postponed until the painting stage.

When the walls and ceilings are primed, the slight imperfections that sometimes occur in finishing drywall will show up. At this time you can have your drywall subcontractor touch up these places so that the walls and ceilings will be ready for final painting (this is called pointing up). Pointing up can also be done just after the interior trim is finished in case the walls get nicked during this process. I have done it *both* times, and at no extra charge.

STEP 16: Interior Trim and Cabinets (1–2 weeks)

Doors, moldings, cabinets, countertops, and shelves are installed at this time. This includes kitchen cabinets, bathroom sink bases and medicine cabinets, and all other built-in cabinets such as bookcases. Cabinets that were ordered after the completion of step 8 should be ready for delivery and installation. The cost of installation should be worked out in advance between you and either your carpenter or supplier (or cabinetmaker if you use shop-built cabinets). Interior trim labor usually includes standard trim and moldings. Any special molding or trim work, or any paneling, should be discussed with your carpenter in the planning stages to determine the extra cost of installation. Your millwork supplier, usually the lumber company, can do a complete material take-off for your trim and discuss costs of extra items.

STEP 17: Painting (2–3 weeks)

You are ready for final painting inside. Your exterior painting can be delayed until this time, too, to save your painter from hopping back and forth from job to job. Work this out when you first discuss painting with your sub. You don't want to leave the exterior trim unpainted or unstained too long, as it may warp or get moldy.

STEP 18: Other Trims (1 day–1 week)

It's time to install vinyl floors and ceramic tiles. Wallpaper can be done at this time or delayed until after you move in. The others can't be delayed, because plumbing cannot be completed until they are.

STEP 19: Trimming Out (1–2 weeks)

It's time for the plumber to finish his work. This is called trimming out or setting fixtures. Some of the fixtures he installs must be wired, so he needs to finish before the electrician can finish. Also, your heating and air-conditioning must be completed before your electrician finishes his work. Next comes your electrician to install switches, receptacles (sometimes called devices), light fixtures, and electrical appliances, such as the oven and range. He will also wire the electrical apparatus that has been installed by your plumber and heating and air-conditioning subs.

STEP 20: Cleanup (2–3 days)

Both inside and outside cleaning can be started. The bulk of outside trash, and an incredible amount of it that comes from inside, can be picked up by truck and hauled away. You can do the inside work. If you don't want to, you can get a professional. Be sure to add the cost to your estimate.

STEP 21: Carpet (3 days–1 week)

Hardwood floors should be finished before the carpet is installed. Allow at least three days for hardwood finishing. No other subs should be working in the house while this work is done. They are expensive floors, and it is well worth a few days of "no trespassing" to have them done right. They should be finished before the carpet is installed because of the sanding required before the floors are stained and sealed with polyurethane. The dust from sanding usually is controlled, but some could permeate the rooms with carpeted floors.

Because of this dust, you may want to finish the floors before step 20, cleaning up. I've done it both ways and found advantages to both.

STEP 22: Driveway (1–3 days)

You should keep heavy trucks off newly laid concrete or asphalt drives for a period of time. Try to time it so that you have received all heavy shipments of materials, hauled off all heavy loads of trash, and don't have any heavy equipment coming to do work. Realize that if you wait until the end with the driveway, you'll at least have a moving van on it. Concrete can support a moving van a week after the concrete is poured, but asphalt can't. If you use asphalt, wait until after you move in to pave. Put down a good stone base on the drive before moving in. My asphalt paving contractor insists on this and puts down the stone base and then redoes the stone (or dresses it up) before paving it with asphalt several days after the moving van leaves.

STEP 23: Landscaping (1–3 days)

This is an item that, depending on the requirements of your lender, can be put off until after you move in. Such a situation may arise due to weather or scheduling problems or lack of money due to other cost overruns at the end of construction. You might get by with just grading and seeding, or mulching disturbed wooded areas.

STEP 24: Final Inspections, Surveys, Loan Closings (1–3 days)

After the completion of the actual house (not the driveway and landscaping), all final inspections from the county or city for building, electrical, mechanical, and plumbing should be made. Also, the lender will make a final inspection of the house at this time and may require driveways and landscaping to be complete before it will disburse the balance of the construction loan and close it out to the permanent mortgage.

When approval is given, your attorney will coordinate the necessary paperwork and schedule the loan closing. The lender will most likely require a final survey to be sure no additional structures or additions have been placed on the lot in violation of deed restrictions or zoning. Generally, your attorney will order this final survey and any other necessary documents required for closing.

The actual loan closing may take only about 15 minutes, depending on the efficiency of the attorney and the lender. One document you must remember to obtain (your attorney or lender will remind you of it) is your insurance policy. It will need to be converted into a homeowner's policy prior to closing. This merely takes a phone call to the agent who issued you your builders' risk (or fire) policy.

STEP 25: Enjoy!

Chapter 9

Additions to an Existing Home

IF YOU ALREADY have a house but would like more room and don't want to move, you can add to your present home. This option would also require less money than building a bigger house, as you already have a house purchased when houses were cheaper to build.

If your mortgage interest rate is low, it is difficult to think of exchanging that rate for higher rates. An addition may be the solution for you. It could be easier than you think. You may already have a head start toward additional space.

There are two basic ways to add to a home. One is to make habitable an existing unfinished area such as a basement, garage, screened porch, breezeway, or attic. This is the least expensive way because you already have a room, foundation, and some or all of the outside walls. The other and more expensive way is to build on to your house. If you consider this, check to see that you have room on your lot without violating required setbacks and that you meet any subdivision or deed restrictions.

Don't add too much value to your home. You don't want to have the most expensive house in the neighborhood. It's a real estate fact of life: The $200,000 house in the $100,000 neighborhood is very difficult to sell.

Whether you finish an existing space or add to your present structure, the procedure will be the same. Start at the beginning of this book and treat the addition as a small home. The only things that will be different from building a house are that you already have the land and the costs will be somewhat different. Don't skip any of the applicable steps I outlined earlier for building a house. You will need an attorney to check restrictions and for the loan, as well as plans, an estimate, and the money to build. Look at the end of Chapter 5 (page 56) before deciding to do your own labor.

How to Estimate Costs

The basic differences in estimating an addition against a complete home will be the elimination of some costs due to the existing structure and the increase of some square foot costs due to the smaller size of the job. Subs may want a few more dollars for their labor costs because they could be earning more for almost the same amount of time.

The general contractor that you won't need would also have charged a greater percentage for profit and overhead. This savings will more than offset the increase in labor costs from subs.

You should obtain bids (quotes), as though you were building a house, and proceed from there.

How to Finance an Addition

Financing an addition is far easier than financing a new home. You can simply get a home equity loan or *second mortgage*. Both loans are based on the equity you have in your home. Your equity is the difference between what the house is worth and what you owe on it. Quite often you don't even need to give a reason to borrow against that equity because the lender is well protected if you should default. The house exists, it is not being built. The amount you can borrow varies from lender to lender, but if you have been in your house for at least three years it is safe to assume you have built up some equity. As a rule of thumb you can borrow up to one-half of your equity. The second mortgage, which is usually a five- to ten-year loan, is less expensive per month and in total payback than refinancing the whole house or building a new house of equal dollar value, especially when interest rates are high.

Just as in planning new home construction, your local lender can sit down with you and discuss your needs.

Glossary

ADJUSTABLE RATE MORTGAGE (ARM). A mortgage in which the interest rate is adjusted periodically up or down, usually once or twice a year.

ANNUAL PERCENTAGE RATE (APR). Everything financed in your mortgage loan package (interest, loan fees, points or other charges) expressed as a percentage of the loan amount. APR is usually slightly above the actual interest rate alone.

APPRAISED VALUE. An estimate of the value of property, such as a house.

BALLOON (PAYMENT) MORTGAGE. Usually a short-term fixed-rate loan, it involves small payments for a certain period of time and one large payment for the remaining amount of the principal at a time specified in the contract.

BREAKER. Electrical circuit breaker. The modern version of the old-time fuse.

BRICK VENEER. Brick used in lieu of siding.

BRIDGE LOAN. A short-term loan to bridge the time between the purchase of one house and the sale of another.

BRIDGING. Small pieces of wood or metal used to brace floor joists.

CARPET UNDERLAYMENT. Either plywood or pressed wood (particleboard) placed over the subfloor to make the floor more solid.

CLEAR TITLE. A title (proof of ownership) of any property (land, auto, house) that is free of liens, mortgages, judgments, or any other encumbrance.

CLOSING. The meeting between the buyer, seller, and lender or their agents where the property and funds legally change hands.

CLOSING COSTS. Costs the buyer must pay at the time of closing in addition to the down payment, including points, mortgage insurance premium, homeowner's insurance, prepayments for property taxes, etc. Closing costs usually average 2 to 4 percent of the loan amount.

COLOR RUN. Materials produced using the same batch of dye, such as bricks, carpet, or paint. Subsequent batches may vary in color.

CONTRACT PRICE. A pre-agreed set price for a service or product.

COURSE. One row of bricks.

CRAWL SPACE. A shallow space under a house or porch or in an attic.

DOWN PAYMENT. Money paid to make up the difference between the purchase price and the mortgage amount.

DRIP CAP. A protective molding, usually metal, to divert water over an exterior surface.

DRY-IN. A stage in construction at which time the building is protected from rain and snow.

DRYWALL. Sheetrock.

EARNEST MONEY. Money given by a buyer to a seller as part of the purchase price to bind a transaction or assure payment (also called a binder).

EQUITY. The difference between what you owe on something and what it is worth. This equals the amount of your house that you currently own, and it can be used as collateral for a home equity loan.

ESCROW. An account held by the lender into which the homeowner pays money for tax or insurance payments and also for earnest deposits held pending loan closing.

EXPANSION JOINT. A joint used to separate blocks or units of concrete to prevent cracking due to expansion as a result of temperature changes.

FASCIA BOARD. The flat, vertical board at the edge of the roof.

FLUE. A channel for the passage of hot gases and smoke.

FOOTING. A mass of concrete below the frost line supporting the foundation or piers of a house.

FRAME CONSTRUCTION. Using wood, as opposed to brick, block, concrete, or steel, for building walls, ceilings, and roofs.

FRAMING. The construction of the skeleton of a house, to include walls, floors, ceiling, and roof.

FROST LINE. The depth in the earth at which the warmth of the earth prevents freezing or the formation of frost.

GOOD FAITH ESTIMATE. A requirement that lenders provide borrowers with settlement service charges the borrower is likely to incur. A good faith estimate must be provided by the lender within three business days of receiving a signed loan application.

HEAT PUMP. An electrical appliance for heating and cooling a building. In the heating phase, it draws heat from the outside air and transfers it to the inside of a house or other building.

HOMEOWNER ASSOCIATION. The fees charged to homeowners either in a condominium or a planned unit development for upkeep of common areas.

INTERIM FINANCING. A short-term loan. It is generally converted to a long-term loan at a later date.

JOIST. One of the horizontal, small timbers ranged parallel and used to support a floor or ceiling.

LOAD-BEARING CAPABILITY. The amount of weight a particular substance can withstand without breaking or bending beyond its design. Used to describe soil, steel, and wood.

LOT SUBORDINATION. A process of buying land by which the owner will take a note in lieu of payment and legally take an interest in that land secondary to a party with primary interest such as a savings and loan.

MANAGER'S CONTRACT. A contract with a general contractor by which the latter agrees to act as a manager to construct your house. Under such a contract, you are able to remain the primary general contractor.

MARKET VALUE. The highest price that a buyer would pay and the lowest price a seller would accept on a property. Market value may be different from the price a property could actually be sold for at a given time.

MOLDING. The interior trim in a house.

MORTAR. A mixture of cement, sand, and water used between bricks or concrete blocks to hold them together.

MORTGAGE INSURANCE. Money paid to insure the mortgage when the down payment is less than 20 percent.

NOTE. A written acknowledgment of a debt, such as a promissory note.

ORIGINATION FEE. The fee charged by a lender to prepare loan documents, make credit checks, inspect and sometimes appraise property; usually computed as a percentage of the face value of the loan.

PANEL BOX. Electrical panel. A metal box containing circuit breakers.

PIER. A vertical structural support. Usually a masonry or metal column used to support the house, porch, or deck.

POINTS (LOAN DISCOUNT POINTS). Prepaid interest assessed at closing by the lender. Each point is equal to 1 percent of the loan amount (e.g., two points on a $100,000 mortgage would cost $2,000). The purpose of paying points is to obtain a lower interest rate on the mortgage.

POLYETHYLENE. Plastic film used in home construction to provide a moisture barrier.

PREFAB FIREPLACE. A fireplace that is made of metal as opposed to solid masonry. It usually has a metal flue.

PREPAYMENT. A privilege in a mortgage that permits the borrower to make payments in advance of the due date.

PREPAYMENT PENALTY. Money charged for an early repayment of debt. Prepayment penalties are allowed in some form (but not necessarily imposed) in many states.

PREWIRE. Wiring for various items such as lighting, telephones, intercoms, burglar alarms, and installed before drywall or paneling is applied to walls.

QUOTE. A guaranteed price in advance.

RECORDING FEE. The fee charged to record legal documents in a place of permanent records, such as a county courthouse.

REINFORCING ROD. A steel rod placed in concrete to increase the strength of the concrete. Also known as rebar.

ROOF PITCH. The slope of the roof.

ROUGH-IN. The installation of wiring, plumbing, or heat ducts in the walls, floors, or ceilings before those walls, floors, or ceilings are covered with drywall, plaster, or paneling.

SAW SERVICE. Temporary electrical service used during construction.

SECOND MORTGAGE. The pledging of property to a lender as security for repayment, but using property that has already been pledged for a loan.

SEPTIC SYSTEM. A means of disposing of sewage in the ground.

SHIM. A thin wedge of wood or metal used to fill in a space.

SOFFIT. The underside of a roof overhang.

SOFFIT VENT. Air vent in the soffit to allow air circulation under the roofing and between the roof rafters. Prevents heat buildup and rotting of the wood members in the roof.

STAKING. Placing stakes in the ground prior to building to show the location of the corners of the house.

STUD. One of the uprights in the framing of a wall, usually spaced on 16-inch or 24-inch centers.

SWEAT EQUITY. Equity created by a purchaser performing work on a property being purchased.

TAKE-OFF. The compilation of a list of materials used for a particular phase of construction, such as the number of bricks or the number and sizes of windows. Also called a schedule of materials, it can often be prepared by a lumberyard or other supplier.

TAMPED. Packed down. Tamping soil prevents later settling, such as might occur underneath concrete.

TAX STAMP. A stamp affixed to a legal document to indicate that a tax has been paid.

TEST BORING. Sample of the soil upon which a structure is to be built to determine what weight the soil is capable of carrying.

TITLE INSURANCE. An insurance policy that protects the owner of real estate from any loss due to defects in his title.

TOPOGRAPHICAL PLAT. A drawing showing the surface features of property.

TRANSIT. A surveying instrument used to measure horizontal angles, levelness, and vertical depth.

UNSECURED LOAN. A loan in which no material possessions are pledged as security for repayment.

WATERPROOFING. Making a foundation impervious to water.

Appendix

The following legal instruments are published as examples. Because of varying state laws, these should not be used by you unless such use is approved by your attorney.

Manager's Construction Contract

1. GENERAL

This contract dated _____ is between _____ (owner) and _____(manager), and provides for supervision of construction by Manager of a residence to be built on Owner's Property at _____, and described as _____. The project is described on drawings dated _____ and specifications dated _____, which documents are a part hereof.

2. SCHEDULE

The project is to start as near as possible to _____, with anticipated completion _____ months from starting date.

3. CONTRACT: FEE AND PAYMENT

    A. Owner agrees to pay Manager a minimum fee of $_____ for the work performed under this contract.

        a. Down payment (due prior to start of work) $_____

        b. Framed $_____

        c. Roof on $_____

        d. Ready for drywall $_____

        e. Trimmed out $_____

        f. Final $_____

B. Payments billed by Manager are due in full within ten (10) days of bill mailing date.

C. Final payment to Manager is due in full upon completion of residence; however, Manager may bill upon "substantial completion" (see Paragraph 11 for the definition of terms) the amount of the final payment less 10 percent of the value of work yet outstanding. In such a case, the amount of the fee withheld will be billed upon completion.

4. GENERAL INTENT OF CONTRACT

It is intended that the owner be in effect his own "General Contractor" and that the Manager provide the Owner with expert guidance and advice and supervision and coordination of trades and material delivery. It is agreed that Manager acts in a professional capacity and simply as agent for Owner, and that as such he shall not assume or incur any pecuniary responsibility to contractor, subcontractors, laborers, or material suppliers. Owner will contract directly with subcontractors and obtain from them their certificates of insurance and releases of liens. Similarly, Owner will open his own accounts with material suppliers and be billed and pay directly for materials supplied. Owner shall ensure that insurance is provided to protect all parties of interest. Owner shall pay all expenses incurred in completing the project, except Manager's overhead as specifically exempted in Paragraph 9. In fulfilling his responsibilities to Owner, Manager shall perform at all times in a manner intended to be beneficial to the interests of the Owner.

5. RESPONSIBILITIES OF MANAGER

General

Manager shall have full responsibility for coordination of trades; ordering materials and scheduling of work; correction of errors and conflicts, if any, in the work, materials, or plans; compliance with applicable codes; judgment as to the adequacy of trades' work to meet standards specified; together with any other function that might reasonably be expected in order to provide Owner with a single source of responsibility for supervision and coordination of work.

Specific Responsibilities

1. Submit to Owner in a timely manner a list of subcontractors and suppliers Manager believes competent to perform the work at competitive prices. Owner may use such recommendations or not at his option.
2. Submit to Owner a list of items requiring Owner's selection, with schedule dates for selection indicated, and recommended sources indicated.
3. Obtain in Owner's name(s) all permits required by governmental authorities.
4. Arrange for all required surveys and site engineering work.
5. Arrange for all the installation of temporary services.
6. Arrange for and supervise clearing and disposal of stumps and brush, and all excavating and grading work.
7. Develop material lists and order all materials in a timely manner from sources designated by Owner.
8. Schedule, coordinate, and supervise the work for all subcontractors designated by Owner.
9. Review, when required by Owner, questionable bills and recommend payment action to Owner.
10. Arrange for common labor for hand digging, grading, and cleanup during construction, and for disposal of construction waste.
11. Supervise the project through completion as defined in Paragraph 11.

6. RESPONSIBILITIES OF OWNER

Owner agrees to:

1. Arrange all financing needed for project so that sufficient funds exist to pay all bills within ten (10) days of their presentation.
2. Select subcontractors and suppliers in a timely manner so as not to delay the work. Establish charge accounts and execute contracts with same, as appropriate, and inform Manager of accounts opened and of Manager's authority in using said accounts.
3. Select items requiring Owner selection and inform Manager of selections and sources on or before date shown on selection list.

4. Inform Manager promptly of any changes desired or other matters affecting schedule so that adjustments can be incorporated in the schedule.

5. Appoint an agent to pay for work and make decisions on Owner's behalf in cases where Owner is unavailable to do so.

6. Assume complete responsibility for any theft and vandalism of Owner's property occurring on the job. Authorize replacement/repairs required in a timely manner.

7. Provide a surety bond for his lender if required.

8. Obtain release of liens documentation as required by Owner's lender.

9. Provide insurance coverage as listed in Paragraph 12.

10. Pay promptly for all work done, materials used, and other services and fees generated in the execution of the project, except as specifically exempted in Paragraph 9.

7. EXCLUSIONS

The following items shown on the drawings and/or specifications are NOT included in this contract, insofar as Manager supervision responsibilities are concerned: (List below)

8. EXTRAS/CHANGES

Manager's fee is based on supervising the project as defined in the drawings and specifications. Should additional supervisory work be required because of Extras or Changes occasioned by Owner, unforeseen site conditions, or governmental authorities, Manager will be paid an additional fee of 15 percent of cost of such work. Since the basic contract fee is a *minimum fee*, no downward adjustment will be made if the scope of work is reduced, unless contract is canceled in accordance with Paragraphs 14 or 15.

9. MANAGER'S FACILITIES

Manager will furnish his own transportation and office facilities for Manager's use in supervising the project at no expense to Owner. Manager shall provide general liability and workmen's compensation insurance coverage for Manager's direct employees only at no cost to Owner.

10. USE OF MANAGER'S ACCOUNTS

Managers may have certain "trade" accounts not available to Owner that Owner may find it to his advantage to utilize. If Manager is billed and pays such accounts from Manager's resources, Owner will reimburse Manager within ten (10) days of receipt of Manager's bill at cost plus 8 percent of such materials/services.

11. PROJECT COMPLETION

A. The project shall be deemed completed when all the terms of this contract have been fulfilled and a Residential Use Permit has been issued.

B. The project shall be deemed "substantially complete" when a Residential Use Permit has been issued and less than Five Hundred Dollars ($500) of work remains to be done.

12. INSURANCE

Owner shall ensure that workmen's compensation and general liability insurance are provided to protect all parties of interest and shall hold Manager harmless from all claims by subcontractors, suppliers and their personnel, and for personnel arranged for by Manager in Owner's behalf, if any.

Owner shall maintain fire and extended coverage insurance sufficient to provide 100 percent coverage of project value at all stages of construction, and Manager shall be named in the policy to insure his interest in the project.

Should Owner or Manager determine that certain subcontractors, laborers, or suppliers are not adequately covered by general liability or workmen's compensation insurance to protect Owner's and/or Manager's interests, Manager may, as agent of Owner, cover said personnel on Manager's policies, and Owner shall reimburse Manager for the premium at cost plus 10 percent.

13. MANAGER'S RIGHT TO TERMINATE CONTRACT

Should the work be stopped by any public authority for a period of thirty (30) days or more through no fault of the Manager, or should work be stopped through act or neglect of Owner for ten (10) days or more, or should Owner fail to pay Manager any payment due within ten (10) days written notice to Owner, Manager may stop work and/or terminate this contract and recover from Owner payment for all work completed as a proration of the total contract sum, plus 25 percent of the fee remaining to be paid if the contract were completed as liquidated damages.

14. OWNER'S RIGHT TO TERMINATE CONTRACT

Should the work be stopped or wrongly prosecuted through act or neglect of Manager for ten (10) days or more, Owner may so notify Manager in writing. If work is not properly resumed within ten (10) days of such notice, Owner may terminate this contract. Upon termination, entire balance then due Manager for that percentage of work then completed, as a proration of the total contract sum, shall be due and payable and all further liabilities of Manager under this contract shall cease. Balance due to Manager shall take into account any additional cost to Owner to complete the house occasioned by manager.

15. MANAGER/OWNER'S LIABILITY FOR COLLECTION EXPENSES

Should Manager or Owner respectively be required to collect funds rightfully due him through legal proceedings, Manager or Owner respectively agrees to pay all costs and reasonable attorney's fees.

16. WARRANTIES AND SERVICE

Manager warrants that he will supervise the construction in accordance with the terms of this contract. No other warranty by manager is implied or exists.

Subcontractors normally warrant their work for one year, and some manufacturers supply yearly warranties on certain of their equipment; such warranties shall run to the Owner and the enforcement of these warranties is in all cases the responsibility of the Owner and not the Manager.

(Manager) _____ (seal) Date:

(Owner) _____ (seal) Date:

(Owner) _____ (seal) Date:

Contract to Build House
(Cost Plus Fee)

Contractor: _____

Owner: _____ Date: _____

Owner is or shall become fee simple owner of a tract or parcel of land known or described as:_____.

Contractor hereby agrees to construct a residence on the above described lot according to the plans and specification identified as: Exhibit A — plans and specifications drawn _____ by _____.

Owner shall pay Contractor for the construction of said house cost of construction and a fee of _____. Cost is estimated in Exhibit B. Each item in Exhibit B is an estimate and is not to be construed as an exact cost.

Owner shall secure/has secured financing for the construction of said house in the amount of cost plus fee, which shall be disbursed by a savings and loan or bank from time to time as construction progresses, subject to a holdback of no more than 10 percent. Owner hereby authorizes Contractor to submit a request for draws in the name of Owner under such loan up to the percentage completion of construction and to accept said draws in partial payment hereof. In addition, it is understood that the Contractor's fee shall be paid in installments by the savings and loan or bank at the time of and as a part of each construction draw as a percentage of completion, so that the entire fee shall be paid at or before the final construction draw.

Contractor shall commence construction as soon as feasible after closing of the construction loan and shall pursue work to a scheduled completion on or before seven months from commencement, except if such completion shall be delayed by unusually unfavorable weather, strikes, natural disasters, unavailability of labor or materials, or changes in the plans or specifications.

Contractor shall build the residence in substantial compliance with the plans and specifications and in a good and workmanlike manner and shall meet all building codes. Contractor shall not be responsible for failure of materials or equipment not Contractor's fault. Except as herein set out, Contractor shall make no representations or warranties with respect to the work to be done hereunder.

Owner shall not occupy the residence and Contractor shall hold the keys until all work has been completed and all monies due Contractor hereunder shall have been paid.

Owner shall not make changes to the plans or specifications until such changes shall be evidenced in writing; the costs, if any, of such changes shall be set out; and the construction lender and Contractor shall have approved such changes. Any additional costs thereof shall be paid in advance, or payment guaranteed in advance for the work being accomplished.

Contractor shall not be obligated to continue work hereunder in the event Owner shall breach any term or condition hereof, or if for any reason the construction lender shall cease making advances under the construction loan upon proper request thereof.

Any additional or special stipulations attached hereto and signed by the parties shall be and are made a part thereof.

Contractor: _____ (seal)

Owner: _____ (seal)

              _____ (seal)

Contract to Build House
(Contract Bid)

Contractor: _____

Owner: _____   Date: _____

Owner is or shall become fee simple owner of a tract or parcel of land known or described as _____

Contractor hereby agrees to construct a residence on the above described lot according to the plans drawn by _____, and the specifications herein attached.

Owner shall pay Contractor for the construction of said house $_____.

Prior to commencement hereunder, Owner shall secure financing for the construction of said house in the amount of $_____, which loan shall be disbursed from time to time as construction progresses, subject to a holdback of no more than 10 percent. Owner hereby authorizes Contractor to submit a request for draws in the name of the Owner from the savings and loan, or similar institution, up to the percentage completion of construction and to accept said draws in partial payment thereof.

Contractor shall commence construction as soon as feasible after closing and shall pursue work to a scheduled completion on or before seven months from commencement, except if such completion shall be delayed by unusually unfavorable weather, strikes, natural disasters, unavailability of labor or materials, or changes in the plans and specifications.

Contractor shall build the residence in substantial compliance with the plans and specifications and in a good and workmanlike manner and shall meet all building code requirements. Contractor shall not be responsible for failure of materials or equipment not Contractor's fault. Except as herein set out, Contractor shall make no representations or warranties with respect to the work to be done hereunder.

Owner shall not occupy the residence and Contractor shall hold the keys until all work has been completed and all monies due Contractor hereunder shall have been paid.

Owner shall not make any changes to the plans and specifications until such changes shall be evidenced in writing; the costs, if any, of such changes shall be set out; and any additional costs thereof shall be paid in advance of the work being accomplished.

Contractor shall not be obligated to continue work hereunder in the event Owner shall breach any term or condition hereof, or if for any reason the construction draws shall cease to be advanced upon proper request thereof.

Any additional or special stipulations attached hereto and signed by the parties shall be and are made a part hereof.

Contractor: _____ (seal)

Owner: _____ (seal)

            _____ (seal)

Description of Materials

❑ Proposed Construction   No._____
                                 (to be inserted by FHA or VA)
❑ Under Construction

Property address_____

City_____ State _____

Mortgagor or Sponsor_____
                        (Name)   (Address)
Contractor or Builder _____
                        (Name)   (Address)

INSTRUCTIONS

1. For additional information on how this form is to be submitted, number of copies, etc., see the instructions applicable to the FHA Application for Mortgage Insurance or VA Request for Determination of Reasonable Value, as the case may be.

2. Describe all materials and equipment to be used, whether or not shown on the drawings, by marking an X in each appropriate check-box and entering the information called for in each space. If space is inadequate, enter "See misc." and describe under item 27 or on an attached sheet. The use of paint containing more than one percent lead by weight is prohibited.

3. Work not specifically described or shown will not be considered unless required, then the minimum acceptable will be assumed. Work exceeding minimum requirements cannot be considered unless specifically described.

4. Include no alternates, or "equal" phrases, or contradictory items. (Consideration of a request for acceptance of substitute materials or equipment is not thereby precluded.)

5. Include signatures required at the end of this form.

6. The construction shall be completed in compliance with the related drawings and specifications, as amended during processing. The specifications include this Description of Materials and the applicable Minimum Property Standards.

1. EXCAVATION:
Bearing soil, type _____

2. FOUNDATIONS:
Footings: concrete mix _____; strength psi   Reinforcing _____
Foundation wall: material _____   Reinforcing _____
Interior foundation wall: material _____ Party foundation wall _____
Columns: material and sizes _____ Piers: material and reinforcing _____
Girders: Material and sizes_____ Sills: material _____
Basement entrance areaway_____ Window areaways _____
Waterproofing _____
Footing drains_____
Termite protection_____
Basementless space: ground cover_____; insulation _____;
    foundation vents_____
Special foundations _____
Additional information: _____
_____

3. CHIMNEYS:
Material_____   Prefabricated _____
                                                                                    (make and size)
Flue lining: material_____ Heater flue size_____
Fireplace flue size _____
Vents *(material and size):* gas or oil heater _____ ;
    water heater _____
Additional information:_____
_____

4. FIREPLACES:
Type: ❑ solid fuel; ❑ gas-burning; ❑ circulator _____
                                                                                                (name and size)
Ash dump and clean-out_____ Fireplace: facing _____ ;
    lining _____ ; hearth _____;
    mantel _____
Additional information:_____
_____

5. EXTERIOR WALLS:

Wood frame: wood, grade, and species _____ ❑ Corner bracing.

Building paper or felt _____ Sheathing _____;

thickness_____; width _____;

❑ solid;   ❑ spaced _____" o.c.;   ❑ diagonal: _____

Siding _____; grade _____;

type _____; size _____; exposure_____"; fastening _____

Shingles _____; grade_____;

type _____; size _____;

exposure _____; fastening _____

Stucco _____; thickness _____",

Lath _____ weight _____lbs.

Masonry veneer _____ Sills _____ Lintels _____ Base flashing _____

Masonry: ❑ solid ❑ faced ❑ stuccoed; total wall thickness _____";

facing thickness_____"; facing material _____

Backup material _____; thickness ____"; bonding _____

Door sills _____Window sills _____ Lintels _____ Base flashing ____

Interior surfaces: damp proofing, _____ coats of _____; furring _____

Additional information:_____

Exterior painting: material _____; number of coats _____

Gable wall construction: ❑ same as main walls ❑ other construction

6. FLOOR FRAMING:

Joists: wood, grade, and species _____; other _____;

bridging _____; anchors _____

Concrete slab: ❑ basement floor ❑ first floor ❑ ground supported

❑ self-supporting mix_____; thickness _____";

reinforcing _____ insulation _____; membrane _____

Fill under slab: material _____ thickness _____"

Additional information:_____

7. SUBFLOORING (describe underflooring for special floors under item 21):

Material: grade and species _____; size _____; type _____

Laid: ❑ first floor ❑ second floor ❑ attic ____ sq. ft. ❑ diagonal

❑ right angles

Additional information: _____

8. FINISH FLOORING: *(Wood only. Describe other finish flooring under item 21)*

location	rooms	grade	species	thickness	width	bldg.	paper	finish

First floor _____

Second floor _____

Attic floor _____ sq. ft. _____

Additional information: _____

9. PARTITION FRAMING:

Studs: wood, grade, and species _____ size and spacing _____ other _____

Additional information: _____

10. CEILING FRAMING:

Joists: wood, grade, and species _____ other _____ bridging _____

Additional information: _____

11. ROOF FRAMING:

Rafters: wood, grade, and species _____

Roof trusses (see detail): grade and species _____

Additional information: _____

12. ROOFING:

Sheathing: wood, grade, and species _____ ❑ solid ❑ spaced ___" o.c.

Roofing _____; grade _____; size _____; type _____;

Underlay _____; weight or thickness_____; size _____; fastening _____

Built-up roofing _____; number of plies _____; surfacing material_____

Flashing: material ___; gauge or weight___; ❑ gravel stops ❑ snow guards

Additional information: _____

13. GUTTERS AND DOWNSPOUTS:

Gutters: material _____; gauge or weight _____; size _____; shape _____

Downspouts: material _____; gauge or weight _____; size _____;

　　shape _____; number _____

Downspouts connected to: ❑ storm sewer ❑ sanitary sewer ❑ dry well

　　❑ splash blocks: material and size _____

Additional information: _____

14. LATH AND PLASTER:

Lath ❑ walls ❑ ceilings: material _____; weight or thickness _____

Plaster: coats _____; finish _____

Drywall ❑ walls ❑ ceilings: material _____; thickness _____; finish _____

Joint treatment _____

15. DECORATING *(paint, wallpaper, etc.)*:

rooms	wall finish material and application	ceiling finish material and application
Kitchen		
Bath		
Other		

Additional information: _____

16. INTERIOR DOORS AND TRIM:

Doors: type _____; material _____; thickness _____

Door trim: type _____; material _____;

Base: type _____; material _____; size _____

Finish: doors _____; trim _____

Other trim *(item, type, and location)*_____

Additional information: _____

17. WINDOWS:

Windows: type _____; make _____; material _____; sash thickness ____

Glass: grade _____; ❑ sash weights ❑ balances, type _____;

    head flashing _____

Trim: type _____; material _____

Paint _____; number coats _____

Weatherstripping: type_____; material _____;

Storm sash, number _____

Screens: ❑ full ❑ half; type _____; number _____;

    screen cloth material _____

Basement windows: type _____; material _____; screens, number ___;

Storm sash, number_____

Special windows _____

Additional information: _____

18. ENTRANCES AND EXTERIOR DETAIL:

Main entrance door: material _____; width _____; thickness ____";

Frame: material_____; thickness ____"

Other entrance doors: material _____; width _____; thickness ____"

Frame: material _____ thickness ____"

Head flashing _____ Weatherstripping: type _____; saddles _____

Screen doors: thickness ____"; number _____ screen cloth material

Storm doors: thickness _____"; number _____

Combination storm and screen doors; thickness ____"; number _____;

    screen cloth material _____

Shutters: ❑ hinged ❑ fixed. Railings _____; Attic louvers_____

Exterior millwork; grade and species _____

Paint _____; number coats _____

Additional information: _____

19. CABINETS AND INTERIOR DETAIL:

Kitchen cabinets, wall units: material _____; lineal feet of shelves ____;

    shelf width ____

    Base units: material _____; countertop _____; edging _____

    Back and end splash ____ Finish of cabinets _____; number coats ____

Medicine cabinets: make _____; model _____

Other cabinets and built-in furniture _____

Additional information: _____

20. STAIRS:

	Treads material thickness	Risers material thickness	Strings material size	Handrail material size	Balusters material size
Basement					
Main					
Attic					

Disappearing: make and model number _____

Additional information: _____

21. SPECIAL FLOORS AND WAINSCOT:

		Wainscot		Floors
location	material, color, border, sizes, gauge, etc.	threshold mat'l	wall base mat'l	underfloor mat'l

Kitchen_____

Bath _____

location	material, color, border, cap, sizes, gauge, etc.	height	height over tub	height in showers (from floor)

Bath _____

Bathroom accessories: ❑ Recessed: material _____; number_____

❑ Attached: material _____; number _____

Additional material: _____

22. PLUMBING

fixture	number location	make	mfr.'s fixture identification no.	size	color

Sink _____

Lavatory _____

Water closet _____

Bathtub _____

❑ Shower over tub _____

❑ Stall shower _____

Laundry trays _____

❑ Curtain rod   ❑ Door   ❑ Shower pan: material _____

Water supply: ❑ public ❑ community system ❑ individual (private) system*

Sewage disposal: ❑ public ❑ community system ❑ individual (private) system*

Show and describe individual system in complete detail in separate drawings and specifications according to requirements.

House drain (inside); ❑ cast iron ❑ tile ❑ other _____

House sewer (outside); ❑ cast iron ❑ tile ❑ other _____

Water piping: ❑ galvanized steel ❑ copper tubing ❑ other _____

Sill cocks, number _____

Domestic water heater: type _____; make and model _____;

Heating capacity _____ gph. 100° rise.

Storage tank: material _____; capacity _____ gallons

Gas service: ❑ utility company ❑ liq. pet. gas ❑ other_____

Gas piping: ❑ cooking ❑ house heating.

Footing drains connected to: ❑ storm sewer ❑ sanitary sewer ❑ dry well

Sump pump; make and model _____ capacity _____; discharges into _____

23. HEATING:

❑ hot water   ❑ steam   ❑ vapor   ❑ one-pipe system   ❑ two-pipe system
❑ radiators   ❑ convectors   ❑ baseboard radiation
    Make and model _____
Radiant panel: ❑ floor   ❑ wall   ❑ ceiling   Panel coil: material _____
❑ circulator   ❑ return pump   Make and model_____; capacity ____ gpm.
Boiler: make and model _____
    output _____ Btuh; net rating _____ Btuh
Additional information_____
Warm air: ❑ gravity   ❑ forced   type of system _____
    Duct material: supply _____; return _____
    Insulation _____ thickness _____ ❑ outside air intake
    Furnace: make and model _____
    Input _____ Btuh; output _____ Btuh
Additional information: _____
❑ Space heater   ❑ floor furnace   ❑ wall heater   input ____ Btuh;
    output _____ Btuh; number units _____
    make, model _____
    Additional information: _____
Controls: make and types _____
Additional information: _____
Fuel: ❑ Coal   ❑ oil   ❑ gas   ❑ liq. pet. gas   ❑ electric   ❑ other_____;
    storage capacity _____
Additional information: _____
Firing equipment furnished separately: ❑ Gas burner, conversion type_____
❑ Stoker: hopper feed _____ ❑ bin feed
    Oil burner: ❑ pressure atomizing _____ ❑ vaporizing _____
    make and model _____ control _____
Additional information: _____
Electric heating system: type _____
    input _____ watts; @ _____ volts; output _____ Btuh.
Additional information:_____
Ventilating equipment: attic fan, make and model _____
    capacity _____ cfm _____
    Kitchen exhaust fan, make and model _____
Other heating, ventilating, or cooling equipment _____

24. Electric Wiring:

Service: ❑ overhead ❑ underground____ Panel: ❑ fuse box ❑ circuit breaker

　　make _____ AMPs _____ no. circuits _____

Wiring: ❑ conduit ❑ armored cable ❑ nonmetallic cable ❑ knob and tube
❑ other _____

Special outlets: ❑ range ❑ water heater ❑ other _____

❑ Doorbell ❑ chimes ❑ push-button locations _____

Additional information: _____

25. Lighting Fixtures:

Total number of fixtures _____

Total allowance for fixtures, typical installation, $_____

Nontypical installation _____

Additional information: _____

26. Insulation:

location	thickness	material, type, and method of installation	vapor barrier
Roof			
Ceiling			
Wall			
Floor			

Hardware (make, material, and finish): _____

_____

_____

_____

Special Equipment: (State material or make, model, and quantity. Include only equipment and appliances that are acceptable by local law, custom and applicable FHA standards. Do not include items that, by established custom, are supplied by occupant and removed when he vacates premises or chattels prohibited by law from becoming realty.)

_____

_____

_____

_____

_____

27. MISCELLANEOUS:
(Describe any main dwelling materials, equipment, or construction items not shown elsewhere; or use to provide additional information where the space provided was inadequate. Always reference by item number to correspond to numbering used on this form.)

_____

_____

_____

_____

_____

_____

Porches:_____

_____

_____

Terraces:_____

_____

_____

Garages:_____

_____

_____

Walks and Driveways:

Driveway: width _____; base material _____"; thickness _____"

Surfacing material _____; thickness _____"

Front walk: width _____; material _____; thickness _____"

Service walk: width _____; material _____; thickness _____"

Steps: material _____; treads_____"; risers _____"

Cheek walls _____

Other On-site Improvements:
(Specify all exterior on-site improvements not described elsewhere, including items such as unusual grading, drainage structures, retaining walls, fence, railings, and accessory structures.)

_____

_____

Landscaping, Planting, and Finish Grading:

Topsoil _____ " thick: ❑ front yard   ❑ side yards

   ❑ rear yard to _____ feet behind main building

Lawns (seeded, sodded, or sprigged): ❑ front yard _____;

   ❑ side yards _____; ❑ rear yard _____

Planting: ❑ as specified and shown on drawings ❑ as follows:

   Shade trees, deciduous, _____ " caliper

   Low-growing trees, deciduous, _____ ' to _____ '

   High-growing shrubs, deciduous, _____ ' to _____ '

   Medium-growing shrubs, deciduous, _____ ' to _____ '

   Low-growing shrubs, deciduous, _____ ' to _____'

   Evergreen trees, _____ ' to _____', B&B

   Evergreen shrubs _____ ' to _____', B&B

   Vines, 2-year _____

Identification: This exhibit shall be identified by the signature of the builder, or sponsor, and/or the proposed mortgagor if the latter is known at the time of application.

Date _____ Signature _____

   _____ Signature _____

FHA Form 2005
VA Form 26-1852

Certificate of Insurance

ALLSTATE INSURANCE COMPANY HOME OFFICE — NORTHBROOK, ILLINOIS

Name and Address of Party to Whom this Certificate is Issued　　　**Name and Address of Insured**

INSURANCE
IN FORCE

Types of Insurance and Hazards	Policy Forms	Limits of Liability	Policy Number	Expiration Date
Workmen's Comp.		STATUTORY*		
Employers' Liability	**Standard**	$_____ PER ACCIDENT (Employer's Liability only) *Applies only in following state(s):		

Automobile Liability

		Bodily Injury	Each	Property Damage
❑ owned only	❑ Basic	$	PERSON	
❑ non-owned only	❑ Comprehensive	$	ACCIDENT	$
❑ hired only	❑ Garage	$	OCCURRENCE	$

Bodily Inj. and Prop. Dam. (single limit)

❑ owned, non-owned, and hired		$	EACH ACCIDENT	
		$	EACH OCCURRENCE	

General Liability

Bodily Injury　Property Damage

❑ Premises — O.L.&T.	❑ Schedule	$	EACH PERSON	
❑ Operations — M.&C.		$	EACH ACCIDENT	$
❑ Elevator	❑ Comprehensive	$	EACH OCCURRENCE	$
			AGGREG.	
❑ Products/Completed Operations		$	PROD. COMP. OPTNS.	$
			AGGREGATE OPERATIONS	$
❑ Protective (Independent Contractors)	❑ Special Multi-Peril		AGGREGATE PROTECTIVE	$
			AGGREGATE CONTRACTUAL	$

Bodily Inj. and Prop. Dam. (single limit)

❑ Endorsed to cover contract between insured and		$	EACH ACCIDENT	
_____		$	EACH OCCURRENCE	
date _____		$	AGGREGATE	

The policies identified above by number are in force on the date indicated below. With respect to a number entered under policy number, the type of insurance shown at its left is in force, but only with respect to such of the hazards, and under such policy forms, for which an "X" is entered, subject, however, to all the terms of the policy having reference thereto. The limits of liability for such insurance are only as shown above. This Certificate of Insurance neither affirmatively nor negatively amends, extends, nor alters the coverage afforded by the policy or policies numbered in this Certificate.

In the event of reduction of coverage or cancellation of said policies, the Allstate Insurance Company will make all reasonable effort to send notices of such reduction or cancellation to the certificate holder at the address shown above.

This certificate is issued as a matter of information only and confers no rights upon the certificate holder.

Date _____ , 20____   By _____
                                        Authorized Representative

U454-16
(8-81)

Reading the Plans

The plan on pages 120 and 121 is labeled "Foundation Plan" at the lower left corner. It shows all of the concrete work associated with this house, including the walls, piers, foundation for fireplace, and slabs for the garage and the patio. The furnace and hot-water heater could be placed here as well, or else in a crawl space or under stairs on the first floor.

The plan on pages 122 and 123 is the floor plan. Study yours and ask yourself specific questions about living in a house that is built to your plan. Will you be happy with only a sliding glass door in the family room? Do you want to provide space for both a dining room for formal dining and a separate dinette, or would you prefer to make other use of one of those spaces? Do you want your garage doors opening to the rear of the house, or would you prefer parking closer to the front entrance?

Look at the detail sheet on pages 124–25. It shows the second floor. Have you always wanted a fireplace in your master bedroom? It probably could be arranged — at extra cost. Will you be happy with the laundry up there (where most dirty clothes end up) or would you prefer it on the first floor?

Study the details on pages 126–27. Note the drawing on the left. It shows 3½ inches of insulation — the width of a common 2 x 4. Are you satisfied with that, or do you live in a cold climate where double that amount of insulation would pay for itself in lower fuel bills? Is the kitchen layout what you want, or are there parts of it that will become irritations when it is used?

And finally, the outside elevations. Are you happy with their appearance? Do you want that outside chimney, or would you prefer to see it inside, where its stored heat would be fed back into the home? Will you feel proud of this house when you drive up to it — or when your friends do?

A thorough study of the plans should give you the feel of living in the house. You'll know where your TV set will be placed, and your library of books. You'll have a place for the holiday decorations, the garden tools, the tools and equipment of your hobbies. Your furniture will fit in the spaces planned for it.

Foundation Plan

Floor Plan

Detail Sheet

WALL SECTION
SCALE 1/2"= 1'-0"

Wall Section

D.W.

13' - 2"

8' - 0"

KITCHEN
CABINET DETAILS
3/8" = 1-0"

36" x 60" MIRROR

36" x 48" MIRROR

36" x 48" MIRROR

5' - 2"

5' - 0"

4' - 2"

POWDER RM

HALL BATH UP

MASTER BATH

Outside Elevations

Right Side   Rear Elevation

Left Side

Front Elevation

Metric Conversion Chart

WHEN THE MEASUREMENT GIVEN IS	TO CONVERT IT TO	MULTIPLY IT BY
inches	centimeters	2.54
feet	meters	0.305
miles	kilometers	1.6
square feet	square meters	0.0929
cubic yards	cubic meters	0.765
pounds	kilograms	0.45
°F	°C	°F − 32 X ⁵/₉

Index

Page numbers in *italics* indicate illustrations.

Additions to existing homes, 88–89
Appliances, 51
Attorneys, 5, 57, 87, 89
   contract review and, 13
   land purchase and, 20

Basements, 15–16, 77
Board of Realtors, 17
Bookkeeping, 68
Boxing. *See* Trim, exterior
Brick
   ordering, 44
   take-offs, 44
   veneer, 76
Bridge loan, 33
Builders' cost, 67
Building code. *See* Code
Building costs. *See* Costs
Building permits. *See* Permits
Building sites. *See* Land
Building, steps in, 69–70. *See also*
   names of specific steps such as
   Framing, Plumbing, etc.
Burglar alarm systems, 83

Cabinets, 50, *127*
Carpentry crew, 6–7
   contract for, sample, 59
   finding subcontractors and ,
     58–59
   masonry contractor and, 76
   take-offs and, 43

Carpet, 86
Certificate of insurance, 61
   sample, 117–18
Chimneys, *71*, 83
Cleanup, 86. *See also* Trash
   removal
Clearing and excavation, 47, 73–74
Clear title, 20, 35
Code, 52, 75, 79
Collar beam, *71*
Communication, importance of, 8
Concrete flatwork, 49. *See also*
   Slabs
Construction loans, 30–31, 67–68
Construction site
   phone service, 8, 81
   toilet, 74
Contract bid, 12
   sample, 104–5
Contractors, professional, 8
   profits, 3–4
   using, 12–13
Contracts
   attorney review of, 13
   for carpenter labor, sample, 59
   contingencies, 17
   with professional contractors,
     12–13
     samples, 96–101
   for subcontractors, 61
     sample, 60
Conventional mortgage, 34

Cost estimate sheet
    sample, 45
    using, 43, 68
Cost plus contract, 12
    sample, 102–3
Costs. *See also* Builders' cost
    breakdowns, 47–51
    cutting, 23, 43
    for home additions, 89
    land purchase and, 21–22
    utilities, 20
    variations in, 43–46
    windshield estimating, 41–42
Crawl space, 76

Decision making, house plans and,
    26–27
Delivery of materials, 67
Description of Materials, sample,
    106–16
Detail sheet, 29, *124–25*
Doors, exterior, 48
Down payments, 36
Downspout, *71*
Drainage, 15–16, 72, 76
Drain, footing, *71*, 75
Driveways, 51, 74, 86–87
Drying-in, 79
Drywall, 50, *71*, 84

Electrical systems
    bids, 50, 61
    costs, 20, 47, 50
    floor plans and, 29
    hookup procedures, 56
    inspections, 55
    rough-ins, 83
    temporary service, 55, 74
    trimming out, 85–86
    wiring, 77, 83
Elevations, 29, *128–29*

Equity, 14, 36
Excavation. *See* Clearing and exca-
    vation

Fair market value, 4
Farmers Service Agency Mortgage,
    34–35
Fascia, *71*
Federal Housing Administration
    Mortgage, 34
Fees, 47
FHA. *See* Federal Housing
    Administration Mortgage
Financing. *See* Loans, Mortgages
Fireplaces, 50
    prefab, 83
Fixed fee contract, 12
Flashing, *71*, 83
Floor coverings, 51, 84, 86
Flooring
    finish, *71*
    hardwood, 84, 86
Floor joists, *82*
Floor plan, 29, *122–23*
Footings, 52, *71*, 75
Foundation plan, 29, *120–21*
Foundations, 47, *71*, 75–77
Frame construction, 16, *82*
    inspections, 55
    supply list, 80
Framing, 79–82, *82*
Framing package, 47

Garages
    doors, 51
    slabs, 79
Gas service
    costs, 20, 47
    hookup procedures, 56
Grade line, *71*
Graduated payment mortgage, 35

Gutters, *71*, 72
Gypsum board. *See* Drywall

Hardwood flooring, 84, 86
Header joist, *71*
Heating and cooling
    contracts, 61
    costs, 49
    floor plans and, 29
    inspections, 55
    rough-in, 83
House plans
    changing, 23–26
    getting, 22
    number of sets, 29
    parts of, 27–29
    reading, 119
    studying, 26–27

Inspection Report and
    Disbursement Schedule,
    sample, 32
Inspections
    by builder, 8–9
    final, 55, 87
    by lender, 31, 87
    procedures, 52–56
    subcontractors and, 62, 64
    types required, 55, 83
Insulation, *24*, 25
    costs, 50
    inspections, 55, 84
    of slabs, 78–79
    utility rates and, 25, 84
Insurance
    builders' risk, 47, 56
    certificate of, sample, 117–18
    homeowners', 87
    subcontractors and, 61
Intercom systems. *See* Stereo and
    intercom systems

Interest only payments, 31

Joists
    floor, *82*
    header, *71*

Labor
    doing your own, 56
    interior trim, 50, 85
    rough costs, 47–48
    union, 62
Land
    buying checklist, 21
    costs, 21–22
    fitting plans to, 16, 27–29
    locating, 14
    lot subordination, 36
    sloping lots, 15–16, 72
    soil testing, 17
    wooded, 17
    zoning, 20
Landscaping, 17, 87
Lenders, 9–10, 36–37
Light fixtures, 51
Load-bearing capabilities, 17
Loan closing, 87
Loans, 9–11
    application form, sample, 38–40
    bridge, 33
    carpentry crew and, 6
    construction, 30–31
    down payments and, 36
    for home additions, 89
    lot subordination and, 36
    permanent, 33, 36
Lots. *See* Land
Lot subordination, 36
Lumber companies, 66
Lumber costs, 47

Manager's contract, 12–13
    sample, 96–101
Masonry contractor, 76
Materials
    at builder's cost, 67
    delivery of, 67
    description of, sample, 106–16
    for framing, 80
Mistakes, 9
Mortgages
    amount of, 37
    application form, sample, 38–40
    qualifying for, 36–37
    second, 89
    types, 34–35

National Association of Realtors,
    17

Organization, 6
Orientation of house, 72–73

Painting, 50, 56, 85
    pointing up, 85–86
    priming walls, 84
Percolation test, 18
Permanent mortgage. *See*
    Mortgages
Permits, 47, 52
    sample, 53–54
Phone service
    at building site, 8, 82
    costs, 20, 47
    prewire, 56, 83
Plans. *See* House plans
Plot plan, 27, *28*
Plumbing, *78*. *See also* Water and
    sewage
    bids, 49, 61
    costs, 49
    inspections, 55, 83

rough-in, 77, 83
    trimming out, 85–86
Pointing up, 84–85
Positioning of house, 72–73
Prewires, 56, 83
Priming walls, 84
Profits, 3–4, 14

Questions, 5

Rafters, *71*
Real estate brokers, 4, 7, 14, 17
Record keeping, 6
References, for subcontractors, 62
R factor, *24*, 25
Ridge, *71*
Roofing, 48–49, *71*, 83
Rough-ins, 77, 83

Savings, potential, 3–4, 14
    construction costs and, 23
    insulation and, 25
    loans and, 36, 37
Schedules, subcontractors and, 62
Second mortgage, 89
Septic systems, *18*–19, 50, 74
Setbacks, 27, 73
Sewage. *See* Septic systems; Water
    and sewage
Sheathing, *71*
Sheetrock. *See* Drywall
Siding, 49, *71*, 83
Sill, *71*
Slabs, 52, 78–*79*
Soffit, *71*
Soil testing, 17
Soil treatment, 77
Solar energy, 73
Sole plate, *71*
Spec sheets, 28
Square footage, 22, 42

Staking, 70–73
Stereo and intercom systems, 83
Stone, 75–76
Studs, *71*
Subcontractors, 5, 58–61, 75
   bids from, 61
   contracts with, 61
      sample, 60
   finding, 6, 56, 59–62
   inspections and, 62–63
   insurance and, 61
   paying, 62–63
   references, 62
   scheduling, 62
   terms of trade, *71*
   working with, 64
Suppliers. *See also* Materials
   builder's cost and, 67
   delivery and, 67
   line of credit, 55, 65
   paying, 67–68
   subcontractors and, 7, 55–56
   take-offs and, 7, 43, 66, 85
   types, 65–66
Surveys, 47, 70–73
   final, 87

Take-offs, 7, 43, 47, 66, 85
Technical knowledge, 4–5
Telephone service. *See* Phone service
Termites, 76, 77
Terms of trade, *71*, *83*, 90–95
Test boring, 17
Time requirements, 7–8
Title. *See* Clear title
Toilet, portable, 74
Tools, contractor's, 26
Top plate, *71*
Topsoil, 73

Trash removal, 74, 86
Tree removal, 73
Trim
   exterior, 83
   interior, 50, 85
   other, 85
Trimming out, 85–86

Underlayment, 84
Union labor, 61
Utilities
   costs, 20, 47, 85–86
   hookup, 56, 74
   inspections, 55

VA. *See* Veterans Administration Mortgage, 34
Vapor barrier, for slabs, 78
Variable-rate mortgage, 35
Veneers, 76, *83*
Veterans Administration Mortgage, 34

Wallpapering, 84, 85
Walls
   outer, *24*, *83*
   section, *126*
Water and sewage, 18–19
   costs, 20, 50
   hookup procedures, 56
   temporary service, 74
Waterproofing, 77
Wells, 17, *19*, 74
Windows, 48, *71*, *81*
Windshield estimating, 41–42
Wiring, electrical, 77
Wooded lots, 17

Zoning, 20

Other Storey Titles You Will Enjoy

Be Your Own Home Renovation Contractor, by Carl Heldmann. Everything you need to know to manage any residential renovation project, including evaluating a structure, estimating costs, and hiring good subcontractors. 176 pages. Paperback. ISBN 1-58017-024-2.

Dream Cottages, by Catherine Tredway. Twenty-five unique cottage plans designed to meet a wide range of needs, locations, and lifestyles. For each cottage, the book includes a full-color rendering, elevation views, floor plans, and photographs of architectural details. 176 pages. Paperback. ISBN 1-58017-372-1.

Small House Designs. This collection of award-winning designs for small houses is accompanied by expert commentary, specific technical data, and beautiful exterior photographs. 208 pages. Paperback. ISBN 0-88266-966-4.

New Compact House Designs. This book includes 27 award-winning designs for single-family homes of less than 1,250 square feet. Each design features a site drawing, floor plans, and elevation and section drawings. 192 pages. Paperback. ISBN 0-88266-666-5.

Build a Classic Timber-Framed House, by Jack A. Sobon. Using actual plans, the author shows readers how to build a classic hall-and-parlor home. 208 pages. Paperback. ISBN 0-88266-841-2.

Timber Frame Construction, by Jack Sobon and Roger Schroeder. Explains the complete basics of timber frame construction in terms the beginner can easily understand. 208 pages. Paperback. ISBN 0-88266-365-8.

Building Small Barns, Sheds & Shelters, by Monte Burch. Provides basic, easy construction methods for building attractive outbuildings, such as a multi-purpose barn, a smokehouse, a woodshed, a root cellar, and housing for animals. 248 pages. Paperback. ISBN 0-88266-245-7.

How to Build Small Barns & Outbuildings, by Monte Burch. This book takes the mystery out of small-scale construction. Projects include a pole barn, a horse barn, a greenhouse, a workshop, a guest house, and more. 288 pages. Paperback. ISBN 0-88266-773-4.

*These and other Storey books are available at your bookstore,
farm store, garden center, or directly from Storey Publishing,
210 MASS MoCA Way, North Adams, MA 01247, or by calling
1-800-441-5700. Or visit our Web site at www.storey.com.*